CHEMISTRY RESEARCH AND APPLICATIONS

PHENOLIC COMPOUNDS

STRUCTURE, USES AND HEALTH BENEFITS

Chemistry Research and Applications

Additional books in this series can be found on Nova's website under the Series tab.

Additional e-books in this series can be found on Nova's website under the eBooks tab.

CHEMISTRY RESEARCH AND APPLICATIONS

PHENOLIC COMPOUNDS

STRUCTURE, USES AND HEALTH BENEFITS

PAAVO OLIVER
AND
ANDRUS VILLEM
EDITORS

nova
science publishers
New York

Library of Congress Cataloging-in-Publication Data

ISBN: 978-1-53612-881-9

Published by Nova Science Publishers, Inc. † New York

Contents

PREFACE

Phenolic Compounds: Structure, Uses and Health Benefits opens with a discussion on phenolic substances such as gallic acid, catechin, chlorogenic acid, caffeic acid, p-coumaric acid, ferulic acid, and quercetin. The most common analytical methods (based on spectrophotometric, chromatographic or electrochemical techniques) for determining phenolic compounds applied to a wide range of sample sources are presented. Additionally, the authors study the high concentrations of bioactive substances in fruit berries in order to determine the link between daily fruit intake and human health. A review of the modern literature on extraction, filtration, and adsorption that may be combined with advanced oxidation treatments to minimalize the environmental impact of the remaining wastes is presented, especially focusing on phenolic compounds recovery from olive mill liquidwastes. Lastly, the authors provide an overview on the antiradical and antioxidizing properties of calix[n]arenes and calix[n]resorcinols as part of a larger discussion on the impact of "preorganization" of antioxidant fragments attached to calix[n]arene and calix[n]resorcinol scaffolds and their intramolecular synergy on antioxidant activity.

Chapter 1 - Phenolic compounds are an important class of substances for human health due to their antioxidant activity which can present anti-carcinogenic, anti-inflammatory, and antimicrobial properties. Some examples of phenolic substances are gallic acid, catechin, chlorogenic acid, caffeic acid, *p*-coumaric acid, ferulic acid, and quercetin. These compounds can be found in many plant parts such as roots, leaves, and flower petals, and also in, peels, seeds, and pulps of fruits that are consumed by humans or used to make pharmaceutical formulations. Although in our diet, we consume foods that are rich in polyphenols with antioxidant properties including teas, wines, honey, edible oils, vegetables, rice, beans and fruits (mango, avocado, guava, apple, tomato, for example), other phenolic molecules like cresols, phenols, isomers of benzenediol and eugenol are important environment pollutants due to high toxicity and persistence in the soils, groundwater, rivers, and seawater. These pollutants can be formed by natural degradation of humic substances; however, industrial processes are the main sources of toxic phenol derivatives. Hence, phenolic compounds are found in many different sampling matrices from foods used for human consumption to samples of soils and surface and ground waters. In this chapter the most common analytical methods developed for determining phenolic compounds applied to a wide range sample sources will be presented. These methods are based on spectrophotometric, chromatographic or electrochemical techniques.

Chapter 2 - Numerous studies have pointed out the presence of high bioactive substance concentration in fruit berries as well as a correlation between daily intake of fruits and human health. The greatest benefit to human health has been attributed to phenolic compounds and vitamin C, due to their antioxidant, anticarcinogenic, antimutagenic, antimicrobial, antiinflammatory and neuroprotective properties. A significant part of antioxidant activity in strawberry and blackberry fruits takes place due to phenolic acids. Apart from phenolic acids, flavonoids also make an important group of secondary metabolites, among which the most common in strawberries and blackberries are anthocyanins.

Many factors affect the chemical composition of berries, the most important of them being the genotype, climatic conditions, agricultural practices, fruit ripeness degree and others. The main objective of this study was to investigate the influence of genotype and type of fertilizer (strawberry), as well as the system of cultivation and climatic conditions (blackberry) on the content of vitamin C and phenolic compound.

The application of biofertilizers predominantly expressed a stimulating influence on most of the phenolic compounds in fruit berries tested. In the rain-shield cultivation of blackberries, higher values for all parameters of the fruit phenolic composition were obtained except for vanillic acid content, in comparison to standard cultivation system.

Changes that have occurred in the nutritional quality of the tested strawberry cultivars indicate the best performance was exhibited by 'Joly', which can be recommended for further promotion and expansion in strawberry growing regions. Taking into consideration the stimulating impact of biofertilizers in strawberries and rain-shield cultivation systems in blackberries on the phenolic compound in fruits, the use of biofertilization and rain-shield can be considered justified in terms of improving the current strawberry and blackberry production technology and obtaining products with positive effects on human health.

Chapter 3 - Olive oil production is an important industry especially in the Mediterranean Sea area where almost 97% of the worldwide olive oil is produced. These industries are generally characterized by small mills spread over large areas. Moreover, olive oil extraction generates large amounts of liquid and solid wastes which are characterized by high toxicity and seasonal character. The most generalized management strategy for these wastes is dewatering and soil spreading. However, the increasing concern regarding the impact of this procedure over the soil and aquifers quality is pushing towards the development of more environmentally friendly approaches.

One important feature of olive mill wastes is the presence of phenolic compounds. In fact, olives are rich in those compounds and a high fraction is lost during the extraction process. Only about 2% of phenolic compounds are retained in the olive oil. Thus up to 98% are incorporated in solid and liquid wastes. Phenolic compounds are related with the bactericide and phytotoxic character of these wastes which difficult their biological treatment.

Phenolic compounds present a high variety of biologic activities such as antioxidant, cardioprotector, anti-inflammatory and chemopreventive. Thus these substances present a high added-value and olive mill wastes seem to be an interesting source. In this context, phenolic compounds recovery will integrate added-value compounds obtained through renewable sources with waste management.

Several recovery strategies are presented in literature involving extraction, filtration and adsorption that may be integrated with advanced oxidation treatments to minimize the environmental impact of the remaining wastes. In this context, this chapter aims to make a short overview on the works published regarding phenolic compounds recovery from olive mill liquid wastes.

Chapter 4 - The literature data on antiradical and antioxidizing properties of calix[n]arenes and calix[n]resorcinols have been summarized. The dependence of these properties of the composition and structure of macrocycles and methods for the modification of calix[n]arenes and calix[n]resorcinols by hindered phenolic fragments has been considered. The effect of "preorganization" of antioxidant fragments attached to calix[n]arene and calix[n]resorcinol scaffolds and their intramolecular synergy on antioxidant activity has been discussed.

In: Phenolic Compounds
Editors: P. Oliver and A. Villem
ISBN: 978-1-53612-881-9

Chapter 1

METHODS OF ANALYSIS FOR PHENOLIC COMPOUNDS

Julio Cesar Bastos Fernandes, PhD
Institute of Educational Sciences – ICED,
Federal University of Western Pará – UFOPA, Santarém, PA, Brazil

ABSTRACT

Phenolic compounds are an important class of substances for human health due to their antioxidant activity which can present anti-carcinogenic, anti-inflammatory, and antimicrobial properties. Some examples of phenolic substances are gallic acid, catechin, chlorogenic acid, caffeic acid, *p*-coumaric acid, ferulic acid, and quercetin. These compounds can be found in many plant parts such as roots, leaves, and flower petals, and also in, peels, seeds, and pulps of fruits that are consumed by humans or used to make pharmaceutical formulations. Although in our diet, we consume foods that are rich in polyphenols with antioxidant properties including teas, wines, honey, edible oils, vegetables, rice, beans and fruits (mango, avocado, guava, apple, tomato, for example), other phenolic molecules like cresols, phenols, isomers of benzenediol and eugenol are important environment pollutants due to high toxicity and persistence in the soils, groundwater, rivers, and seawater. These pollutants can be formed by natural degradation of humic substances; however, industrial processes are the main sources of toxic phenol derivatives. Hence, phenolic compounds are found in many different sampling matrices from foods used for human consumption to samples of soils and surface and ground waters. In this chapter the most common analytical methods developed for determining phenolic compounds applied to a wide range sample sources will be presented. These methods are based on spectrophotometric, chromatographic or electrochemical techniques.

Overview of Spectroscopy Methods

Many spectroscopic methods are based on absorption, emission, reflectance, and scattering of visible, ultraviolet, and infrared electromagnetic radiation. However, energetic particles such as ions or electrons from X-rays, microwaves, and radio waves electromagnetic radiation sources are also employed in spectroscopy. In each case, the method is named according to the interaction form of the electromagnetic radiation or energetic particles with the matter [1].

Spectroscopy analysis started around 1860 with the work of Gustav Kirchoff and Robert Bunsen [2]. Throughout the modern history of scientific exploration, spectroscopy techniques have been fundamental for the discovery of many chemical elements. Advances in these techniques include Fourier transform infrared spectroscopy that is based on the interferometer, a device invented by Albert A. Michelson in 1892 [3]. Pieter Zeeman discovered the effect of double lines in the sodium spectrum when a magnetic field was applied to the flame containing this metal [3]. This effect is used to reduce the chemical interferences in atomic absorption spectrometry [4]. Bohr's atomic theory was a milestone for the development of spectroscopy because it theoretically predicted the emission lines in the hydrogen spectrum [5]. In 1928, C. V. Raman and K. S. Krishnan experimentally observed a new type of secondary radiation with a different wavelength of incident light, which was scattered by the molecules in a transparent medium [3, 6]. Another milestone for spectroscopy was the development of spectrophotometers by Arnold O. Beckman that allowed meaningful growth of the field of biochemistry [7, 8].

Currently, spectroscopic techniques are powerful tools for chemistry, space studies, physics, optics and other fields of science. These techniques are also currently used in the analysis of total phenols in diverse samples wherein the majority of these techniques are based on light absorption in the ultraviolet, visible and near-infrared region.

MOLECULAR SPECTROPHOTOMETRY FOR PHENOLS ANALYSIS

The most widely-used spectrophotometric method for analysis of phenolic compounds based on light absorption is the Folin-Ciocalteau method [9]. This method was developed in 1927 in order to overcome problems with formation of colored compounds with phenols found in chromophores using phosphotungstic and phosphomolybdic acids [10, 11]. The range of proportionality obtainable for determining different amounts of tyrosine and tryptophan using phosphotungstic and phosphomolybdic acids was limited because the reaction between these amino acids and the phenol reagent was not complete, and the reagents themselves presented a characteristic blue color [9], and also because the active ingredients in the phenol reagent are unstable since they are decomposed in basic conditions; however, the phenol reagent reacts only with tyrosine in alkaline conditions [9]. Therefore, a large excess of the phenol reagent was necessary to get the complete reaction with tyrosine or tryptophan to obtain maximum color intensity. One of the drawbacks in using large quantities of phenol reagent is that it produces turbidity or precipitates sodium salts [9]. Folin and Ciocalteau proposed to resolve these problems preparing the phenol reagent under reflux during 10 h in medium containing a higher concentration of phosphoric and hydrochloric acids plus lithium sulfate to increase the solubility of the molybdic and tungstic complexes [9, 12].

Therefore, the principle of Folin-Ciocalteau method is based on the reduction of both Mo (VI) and W (VI) to Mo (V) and W (V) by reducing agents (Figure 1) [13-15]. The complex formed absorbs between 750 nm and 770 nm and depends on the analyte [12, 13, 16, 17]. For complete color development it is necessary to wait between 30 to 120 minutes; however, the maximum wavelength absorption is reduced with time [9, 12, 16].

The anion derivatives of phosphotungstic and phosphomolybdic acids have α-Keggin structure and the blue complex has a big wheel structure of Mo154-type cluster (Figure 2) [18-21].

Normally, the analysis by the Folin-Ciocalteau method involves the following steps: phenol reagent is added to samples together with deionized or distilled water; the mixture is shaken, and then 15% sodium carbonate is poured into the mixture [16]. The order of reagent addition is very important in this spectrophotometric method. A white precipitate is observed when phenol reagent is added to the alkaline medium without previously diluting the sample. The precipitate is normally attributed to the sodium salts of molybdenum and tungsten complexes [12, 22].

The Folin-Ciocalteau method has been used extensively for determining total phenols in many different samples, as can be seen in Table 1. Although this method is employed in a direct assay or as a reference method for analysis of phenolic compounds, phenol reagent used with phosphotungstic and phosphomolybdic acids is limited. The reagent reacts with any reducing compound such as vitamins (ascorbic and retinoic acids), thiols (glutathione), amino acids (tyrosine, tryptophan and cysteine); proteins, nucleotide bases (guanine), unsaturated fatty acids (arachidonic acid - slight reactivity), carbohydrates (glyceraldehyde and dihydroxyacetone), inorganic ions (iron II, manganese II, iodide, sulfite), ketones (butanedione) and metal complexes. Copper complexes increase the reactivity of salicylate derivatives with phenol reagent. Thus, the Folin-Ciocalteau method should be described as a measurement of the total antioxidant capacity of a sample. In plants, phenolic compounds are the main antioxidants, and frequently, the Folin-Ciocalteau method is used to quantify the total content of phenolic compounds with reasonable accuracy [17].

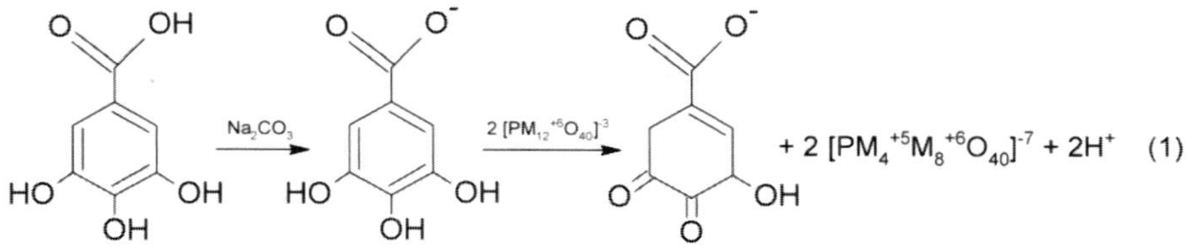

M = Mo or W

Figure 1. Reaction between phenolic compounds and derivatives of phosphotungstic and phosphomolybdic acids in alkaline medium resulting in the formation of a blue color in the Folin-Ciocalteau method.

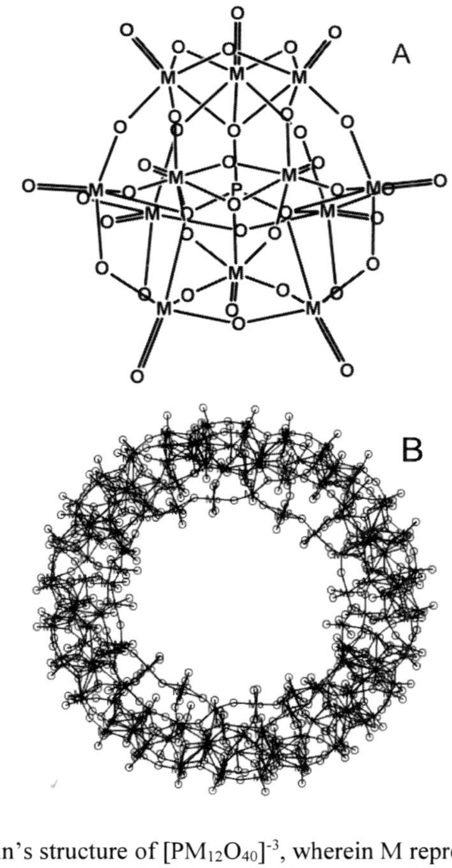

Figure 2. A. α-Keggin's structure of $[PM_{12}O_{40}]^{-3}$, wherein M represents either molybdenum (Mo) or tungsten (W). B. Big wheel structure of the molybdenum blue complex $[Mo_{126}{}^{+6}Mo_{28}{}^{+5}O_{462}H_{14}(H_2O)_{70}]^{14-}$.

Table 1. Analysis of total phenols using the Folin-Ciocalteau method

Samples			Gallic acid concentration ($mg.kg^{-1}$)	Reference
Açaí berry pulp			54-76	[23]
Apple	*Malus niedzwetzkyana*	Peel	2.06-4.73	[24]
		Flesh	0.34-1.06	
	Malus domestica	Peel	1.64-.290	
		Flesh	0.16--.29	
Avocado seeds			35-47	[25]
			0.88-2.75	[26]
Beer	Larges		0.03-0.18	[27]
	Ales/porters/stouts		0.19-0.41	
	low calorie		0.03-0.09	
	Non-alcoholic		0.03-0.07	
Biowaste and biowaste additives:				[28]
Sewage sludges, animal fertilizer, sawdust, palm bark.			0.07-0.15	
Mixture containing organic and mineral matter			0.10–0.80	
Provence cane, leaves, bagasse from sugar cane, shoots.			0.48–4.83	
Chocolate			2.6-48.5	[29]
Cork in boiling water			1.9-6.0	[30]
Guava	Fruit		7.2-7.8	[31, 32]
			70-115	[32, 33]
			16-18	[32, 34]
			145-163	[32, 35]
			170-345	[32, 36]
			1264-2473	[32, 37]
	Leaves		159-175	[32, 38]
			267-483	[32, 33]
Honey	Rape		0.39-0.69	[39]
	Sunflower		0.40-0.52	
	Lime		0.34-0.40	
	Acacia		0.43-0.58	
	Heather		1.16-1.51	
	Cherry		0.99-1.10	
	Polifloral		0.72-1.34	
	Raspberry		1.86-1.93	
	Fir		1.68-2.01	
	Wild flowers		0.43-0.44	[40]
	Eucalypt		0.38-0.40	
	Orange		0.16-0.18	
Honeydew			1.34-1.81	[39]

Mediterranean plant: *Satureja montana subsp. kitaibelii*		23-29	[41]
Mung bean	Black	8.6-9.4	[42]
	Green	11.3-11.7	
Olive oil		0.23-0.35 0.29-2.18 0.12-0.20	[43] [44] [45]
Propolis		0.66-0.81 0.03-0.33	[40] [46]
Rice	Black	2.53-7.09	[47]
	Red	2.66-4.94	
Teas	Black	26-28 76-117 29-83	[48] [49] [50]
	Green	78-140 82-144 66-68	[49] [50] [51]
	Bamboo leaf	11-15	[50]
	Lemongrass	13-17	
	Lotus leaf	20-32	
	Mulberry leaf	11-13	
	Rosemary	30-39	
Teas	Rooibos	16-39	[50]
	Peppermint	33-78	
	Persimmon leaf	14-47	
	Mate	27-67 30-32 11-80	[50] [51] [52]
	Oolong	51-52	[51]
	Red Lapacho	4	[49]
	White	49-100	[49]
	Ficus sycamore	139-141	[53]
	Lippia javanica	120-130	
	*Aspalathus linearis (*Rooibos)	16-39 58-75	[50] [53]
	Myrothamnus flabellifolius	44-51	[53]
	Fadogia ancylantha	24-34	[53]
	Adansonia digitata	5-11	[53]
	Centella asiatica	7-14	[54]
Teff	Flour	1.74-1.80	[55]
	Brown	1.86-2.19	
	White	1.39-1.44	

Table 1. (Continued)

Samples		Gallic acid concentration ($mg.kg^{-1}$)	Reference
Wines	Red	2-3	[56, 57]
	Chardonnay	0.23-0.37	[58]
		0.018-0.034	[59]
	Chenin blanc	0.021-0.022	[59]
	Sauvignon blanc	0.025-0.026	
	Cooking	0.024-0.025	
	Raspberry	1.42-1.46	[60]
	Blackberry	2.23-2.33	
	Blueberry	2.23-2.29	
	Black chokeberry	2.33-2.41	
	Apple	0.58-0.77	
	Sour cherry	1.90-2.18	
Red wine pomace		25-43	[61]

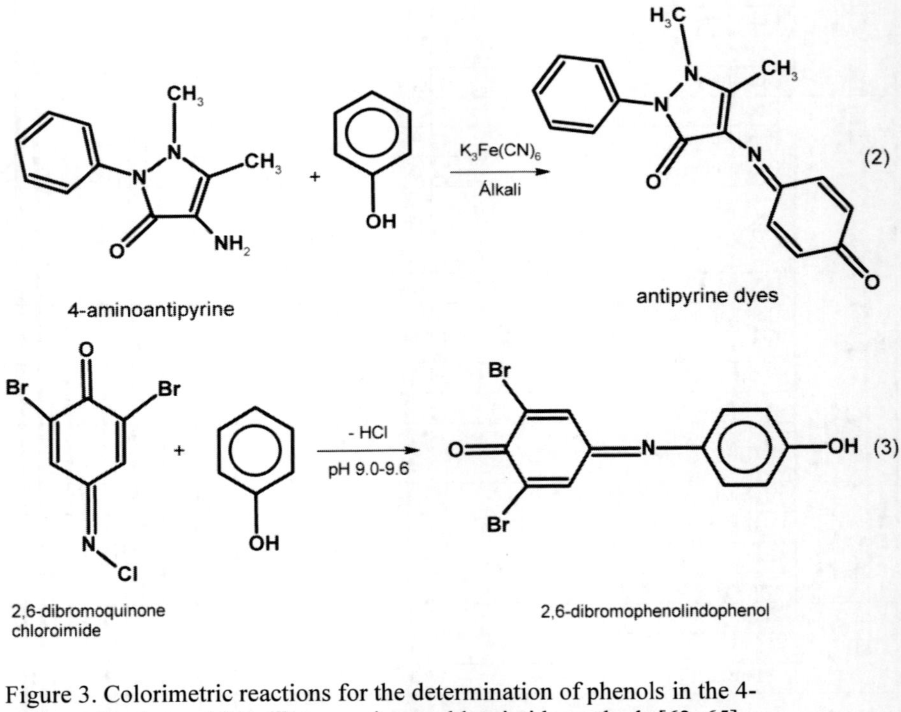

Figure 3. Colorimetric reactions for the determination of phenols in the 4-aminoantipyrine and 2,6-dibromoquinone chloroimide methods [63, 65].

Other spectrophotometric methods have been developed to overcome the problems inherent in the Folin-Ciocalteau method. Two colorimetric reactions have also been described for determining phenolic compounds. The first reaction uses 4-aminoantipyrine to form a red dye that absorbs at 510 nm. The second one employs 2,6-dibromoquinone chloroimide (Figure 3) to produce a blue compound which exhibits maximum absorption at 610 nm [63, 64].

Both reactions are favorable with phenols containing a free *p*-position. If this position in phenol compounds is occupied by alkyl, aryl, ester, nitro, benzoyl or aldehyde groups, no reaction occurs. Quinonechloroimines also react with aromatic amines, uric acid and thiouracil derivatives [65, 66]. However, the main disadvantage of these methods is the time required for color development and the signal intensity which is greatly influenced by the pH [62-64, 67].

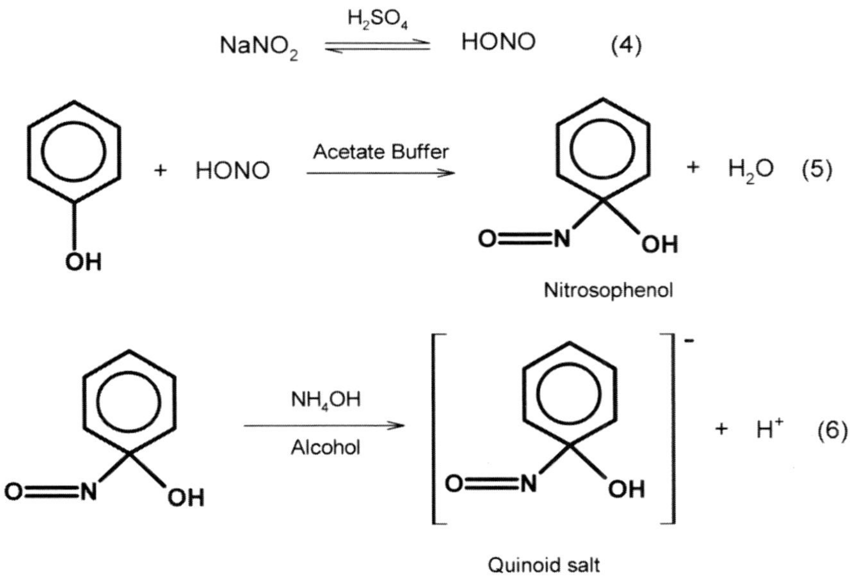

Figure 4. Colorimetric reaction for determination of phenols in the nitrosophenol method [67].

The nitrosophenol method is another spectrophotometric approach to measure phenol compounds. This method allows determination of *ortho-*, *meta-* and *para-*substituted phenolic substances, and is based on the reaction between nitrous acid and phenol to form a nitrosophenol compound. This compound is rearranged to generate a colored quinoid salt in alkaline and alcoholic medium that absorbs at 420 nm, and nitrous acid is produced *in loco* using sodium nitrite and sulfuric acid in acetate buffer. The scheme of this reaction is presented in Figure 4 [67].

The main disadvantage of the nitrosophenol method is the presence of inorganic salts dissolved in water and phenolic molecules themselves, which provoke changes in the intensity of the dye color [67].

Although colorimetric methods are very sensitive, in order to detect levels of phenolic compounds at parts per billion (ppb), ultraviolet absorption has been employed. Fountaine et al. devised a method based on two hollow cathode lamps, a platinum lamp to detect ultraviolet lines at 289.4, 289.8 and 293.0 nm, and a chromium lamp to detect visible lines at 357.9, 359.3 and 360.2 nm. When an acid solution of the phenolic compound changes to alkaline, the ultraviolet radiation is absorbed while visible radiation does not undergo absorption. Therefore, monitoring the ultraviolet bathochromic shift was suitable to determine the amount of phenolic compounds at the ppb level. The advantage of this procedure was its capability to determine *p*-blocked phenols, although the instrumental system was very complex [68].

The emergence of dynamic methods such as flow injection analysis (FIA) allowed for a reduction in the consumption of reagents and a decrease in the time required to determine phenolic compounds by spectrophotometry. Frenzel et al. developed a flow injection procedure using 3-methyl-2-benzothiazoline hydrazone (MBTH) as a chromophore reagent. This method was very sensitive for phenol substances, reaching detection levels between 12 and 30 ppb, and the sampling rate using FIA-MBTH was about 60 samples per hour. These authors commented that despite the accuracy of this method to determine *p*-substituted phenols, the adequate choice of the maximum wavelength for phenol compounds was a problem, because the colored products presented low absorption between

410 and 530 nm [69]. Another flow system for spectrophotometric determination of phenols was based on the 4-aminoantipyrine method. This system allowed a sampling rate of 90 samples per hour and reagent consumption that was 200-fold lower than the conventional procedure. The method had sensitivity at the ppb level using a flow cell with a 100 cm optical path that was 80-fold more sensitive than flow cell using an optical path of 1 cm [70]. Nevertheless, the 4-aminoantipyrine method is limited to determining phenols containing a free *p*-position or phenolic compounds *p*-substituted by aryl, alkyl, nitro, benzoyl and aldehyde groups [69].

So far, spectrophotometric methods are not very selective for analysis of phenolic compounds and some cases have the same limitations as previously described in this chapter. Through univariate measurements, that is, measurements using only the maximum wavelength, it is impossible to determine a specific phenolic compound in a complex sample of organic material such as food used for human consumption. In the last few decades, advances in computational methods have allowed for multiple determinations of chemical species with minimum sample preparation using multivariate calibration or chemometrics methods. These methods consist of relating many measurement variables to the quantification of a target variable [71]. Two chemometric methods that are widely used in chemistry are the classical least square (CLS) and partial least square (PLS) methods. CLS can be expressed by the following matrix equation [72]:

$$X = 1\alpha + YA + E \quad (1)$$

Where

X is the $n \times k$ X-block, matrix of absorbance for sample;

Y is the $n \times m$ Y-block, matrix of concentration for pure constituents;

1 is a $n \times 1$ vector of ones;

$\alpha = [\alpha_1, \ldots, \alpha_k]$ is the $1 \times k$ row vector of offsets;

$A = \{a_{lj}\}$ is the $m \times k$ matrix of absorbance for pure constituents;

$E = \{e_{ij}\}$ is the $n \times k$ matrix of random noise terms.

α, A and the variance of the random noise are unknown parameters to be estimated from the data.

PLS is a statistical model that is more efficient for chemical analysis of multiple analytes than CLS, since the use of PLS does not require complete information about interference [72]. There are many PLS algorithms, and among them the principal ones are PLS1 and PLS2. In the former, the regression is done for one dependent variable at a time, while for PLS2 the complete set of dependent variables is computed simultaneously. PLS2 is very useful where the instrumental response is related to many different conditions such as pH, concentration, density and viscosity that can be associated with an infrared spectrum [73].

A simplified model of PLS may be described by computing the score of a regression of two matrices chains: X and Y. Scores are the projections of the sample points on the direction of the main component. PLS is similar to principal components regression (PCR), but in this last case only the X block is designed in space [74-76].

PLS can be presented in the form of two equations [77]:

$$X = TP + E \tag{2}$$

$$Y = TQ + F \tag{3}$$

Where X represents the experimental measurements and Y the concentrations; T is the score, being common to both equations; P and Q are loadings, that is, the cosine of the angle of the direction vector. E and F are matrices of errors for the X and Y blocks, respectively.

The SIMPLS algorithm is an alternative approach to PLS. In SIMPLS scores are calculated as a linear combination of the original variables, in order to maximize a covariance criterion (numerical interrelation between two random variables). The advantage presented by SIMPLS is a lower computer memory requirement since there is no need for reducing the dataset, which results in faster processing [78].

Chemometrics is very important in food analysis, with use of near infrared (NIR) quite common in the food industry [78]. Polyphenolic compounds such as trans-resveratrol, oenin, malvin, catechin, epicatechin, quercetin and syringic acid were determined in red wines by UV–VIS–NIR spectroscopy using PLS model. In this case, high performance liquid chromatography (HPLC) was used as a reference method to measure these phenolic compounds in the wine [79]. NIR-PLS was also applied to the quantification of methylxanthines and phenolic compounds in yerba mate [52, 80]. Methylxanthines have pharmacological properties since they stimulate the central nervous system and the myocardium, and methylxanthines, polyphenols and saponins are phytochemical compounds with great health benefits [80]. HPLC and the Folin-Ciocalteau assay were employed as reference methods for determining methylxanthines and total phenols, respectively [52, 80].

In order to verify the quality of organic compost, phenolic compounds were used as a parameter to manage the changes organic matter changes in compost used in agriculture. In this work, the researchers used middle and near infrared to determine total phenolic compounds based on models built through PLS regression, with the Folin-Ciocalteau method used as a reference to measure total phenols in the compost samples [28].

Another chemometric method that uses attenuated total reflectance (ATR)-Fourier transformed-infrared (FT-IR) spectroscopy was employed to determine antioxidant capacity and total phenols in chocolate samples containing different amounts of cacao. Antioxidant capacities were quantified with DPPH (2, 2-diphenyl-1-picrylhydrazyl) and ORAC (oxygen radical absorbance capacity) assays while procyanidins monomers and phenolic substances were measured by high-performance liquid chromatography coupled with photodiode array detector (HPLC–DAD) and the Folin–Ciocalteu assay, respectively. PLS models were constructed for determining the linear relationship between FT-IR spectra information and the reference values, but it was not feasible to establish a good correlation between cacao content and the factors measured. Nevertheless, antioxidant capacity and total phenols of chocolate can be predicted by the

PLS model. Thus, compared to the conventional chemical assays, FT-IR-PLS reduced the time of analysis from hours to minutes, although it is still important to develop other robust and precise prediction models [29].

Phenolic compounds were also measured accurately in mung bean grains through NIR calibration curves using HPLC as a reference method with the PLSR model in the wavelength range from 1600 to 2500 nm [81].

As described previously, PLS models associated with spectroscopy methods have been showing promising results to efficiently and relatively quickly determine phenolic compounds in food samples, but the construction of the chemometrics model still depends on classical assays that sometimes are not very accurate.

OVERVIEW OF CHROMATOGRAPHIC METHODS

The history of chromatography is a complex subject, but it is well established that the term was first mentioned by the botanist Mikhail Semyonovich Tswett in 1906. In his work, Tswett separated plant pigments using a chromatographic column containing calcium carbonate and therefore the invention of this technique has been attributed to him [82].

Chromatography is a physical method of separation wherein the sample components are distributed into two phases, one stationary and another mobile. Chromatographic techniques are associated with spectrophotometry or electroanalytical methods, among others, to register analytes present in the samples. Many chromatographic methods have been created for the determination of phenolic compounds in foods. In these cases, the analyses are considered to be more accurate than those based on spectrophotometric methods.

The development of chromatography was stimulated by the research of Martin and Synge who worked at the Wool Industry Research Association in Leeds, UK [83]. In 1941, they published the first article about liquid-liquid partition chromatography [84]. In 1952, Martin and James reported their studies about gaseous chromatography for separation of eleven volatile weak acids. In this work, Martin and James employed titration with sodium hydroxide solution using a phenol red as an indicator [85].

Later, other chromatographic techniques were developed. Klesper et al. reported on high-pressure gas chromatography, which also became known as supercritical fluid chromatography because it uses gas above its critical pressure and temperature, which causes an increase in its density, similar to that of liquids [86, 87]. In the next decade a rapid advancement occurred with the development of high-performance liquid chromatography (HPLC), based on ideas that liquid chromatography could be more efficient by reducing the packing particle diameter below 150 mm together with the application of high pressure to increase the speed of the mobile phase [88].

CHROMATOGRAPHY FOR ANALYSIS OF PHENOLIC COMPOUNDS

Chromatographic analysis of phenolic compounds in foods frequently employs liquid chromatography (LC), since phenolic compounds present polar features and low volatility. Phenolic substances are measured by LC using both ultraviolet detectors (LC-UV) or coupling a mass spectrometer (LC–MS). It is also common to use liquid-liquid and solid phase extraction to obtain aqueous samples for determining phenolic compounds by LC-UV. In this case, due to the higher detection limit, there is a need to pre-concentrate aqueous samples [89].

Based on this idea, López-Cobo et al. developed a chromatographic methodology for the determination of phenolic compounds in avocado peel, pulp, and seed. A solid-liquid extraction was used to obtain the polar fraction of the samples. Avocado pulp presented a concentration of phenolic compounds higher than avocado peel and seed, but the peel and seed from avocado contained more polar compounds. The avocado samples were analyzed using HPLC coupled with a diode-array detector (DAD) and electrospray ionization/quadrupole-time-of-flight high-definition mass spectrometry (ESIQTOF-MS). The compounds in the chromatogram were identified comparing the relative values of retention time, UV-VIS spectra between 200 and 600 nm and mass spectra obtained with the information described in the literature. Therefore, the amount of reagent needed to create the calibration curve was reduced. The mobile phase consisted of water containing 1% acetic acid as the solvent in the system A and acetonitrile as solvent in the system B, while in the stationary phase a Poroshell 120 EC-C18 (4.6 mm × 100 mm, silica particle size: 2.7 μm) was used [90]. A similar technique (Liquid chromatography coupled to electrospray ionization and tandem mass spectrometry - LC-ESI-MS/MS) was employed to analyze phenolic compounds in honeydew honey. The chromatographic column used was the VENUSIL C18 (100 mm × 2.1 mm, silica particle diameter: 3 μm) and the mobile phase was composed of 0.1% formic acid in water (system A) and acetonitrile (system B). Among phenolic substances analyzed, 3,4-dihydroxybenzoic acid, benzoic acid, and salicylic acid were the compounds found in higher content in honeydew honey [91].

Wu et al. developed a solid-phase extraction method for phenolic compounds based on magnetic carboxylated multi-walled carbon nanotubes (c-MWCNT-MNPs). This extraction method was associated with liquid chromatography–tandem mass spectrometry (HPLC-MS/MS) for determining phenolic substances in sesame oil. In the HPLC-MS/MS system a C_{18} column (Hypersil Gold, 100 mm × 2.1 mm, particle diameter: 3 μm) was used. For separation of the phenolic compounds the researchers used a gradient elution of binary solvent system, that consisted of 0.01% v/v acetic acid in methanol (system A) and 0.01% v/v acetic acid in water (system B). The HPLC analysis of sesame oil showed low amounts of phenyl alcohols and flavonoids and high contents of lignans and phenyl acids. Among the main phenyl acids that they identified were ferulic, vanillic, syringic and gallic acids [92].

Phenolic compounds were also identified in red, brown and white sorghum whole grain using HPLC-ESI-MS [93]. In this case, the mobile phase was the same described by Seraglio et al. [91]. A large range of phenol classes was found in the hydromethanolic extracts such as free phenolic acids, flavonoids, phenylpropane glycerides, and phenolamides. The brown sorghum extract showed higher diversity and amount of phenolic compounds than red or white sorghum [93].

Another important application of chromatographic analysis in food science is to determine free and bound phenolic compounds. Generally, bound phenolic compounds are esters due to the reaction between phenolic acids and hydroxyl groups from proteins of the cell wall. Esters linked to cell walls are polymers, consisting mainly of monomers of ferulic acid that can dimerize to generate bridges of diferulic acid cross-linking arabinoxylan chains. Free phenolic compounds are commonly proanthocyanidins or flavonoids [16, 94]. Therefore, a diet rich in fiber presents high antioxidant activity.

Sumczynski et al. measured the content, free and bound phenols, in black and red rice by HPLC-DAD. The analytical separation of these compounds was carried out using a C_{18} column (150 mm × 4.6 mm, particle size: 2.6 μm). The samples were eluted through the column using a mobile phase of two systems (A: 1% aqueous solution of acetic acid and B: aqueous mixture of acetonitrile and acetic acid in ratio 32:1) [47]. Free phenols were extracted from rice samples using an aqueous solution of methanol at 80% in an ultrasonic bath at 35°C for 1 h as described by Kotásková et al. [55]. After centrifugation, the pH of the supernatant containing the free phenols was adjusted to 4.5-5.5, and the precipitate obtained by centrifugation of the methanolic extract that was previously used to separate free phenols was washed with water and bound phenols were extracted with 4 M NaOH for 2 h in an ultrasonic bath. The pH of this mixture was also adjusted to 4.5-5.5. Subsequently the alkaline extract was centrifuged and the supernatant then contained the bound phenols. The phenolic compound found in higher amount bound in both rice samples was ferulic acid, followed by quercetin in red rice and vanillic acid in black rice [47]. The same protocol, with small modifications, was employed by Kotásková et al. for determining free and bound phenols in teff samples. In this case, the pH of the extracts was adjusted to 4.0-4.5, and bound phenolic compounds were extracted with 0.4 M NaOH aqueous solution [55]. The main free phenolic compound in brown teff was protocatechuic acid while in white teff it was rutin. Once again, ferulic acid was the bound compound present in higher amount in the sample [55].

Another important application of chromatography in food analysis is to determine variation in phenol concentration with storage time and temperature conditions. Álvarez-Fernández et al. studied this behavior in beverage manufactured from strawberry by fermentation [95]. These researchers found that ellagic acid hexoside, ellagic acid, *p*-coumaric acid, *p*-coumaroyl hexose, catechin, and procyanidin B1 showed higher concentrations in the initial samples, but the content of the last three compounds reduced with storage time. Furthermore, *p*-coumaric acid showed an increase during storage time that was not proportional to the decrease of concentration in p-coumaroyl hexose. This phenomenon is due to disappearance of coumaroyl anthocyanins during the aging process [95, 96].

Table 2 shows some phenolic substances present in different food samples. In this table, the concentration value represents the average between the minimum and maximum value found in the food, independent of the condition of the molecule being free or bound to the cell wall, type of food, and in the case of fruits, if the analysis was performed in pulp, seed or peel.

The phenolic profile in food has been determined mainly by high-performance liquid chromatography associated with mass spectrometry. Although there are changes in the stationary and mobile phases, these changes were made to improve the separation efficiency. Therefore, there have not been many new discoveries using HPLC for analysis of phenolic compounds, and most research efforts have been focused on improving the extraction methods. Supercritical fluid has also been suggested for improving the extraction of phenols in food. Supercritical CO_2 was employed in the extraction of phenolic compounds in açaí (*Euterpe oleracea*) berry oil. The phenol content in the samples was determined before and after the use of carbon dioxide in supercritical conditions and through this method the phenol extracted increased by 37% (55.2 to 75.7 mg g^{-1} of the sample) [23].

Table 2. Phenolic profile of some foods

enolic npounds	Foods (μg g⁻¹ matter) [References]													
	Avocado [90]	Fruit wines							Guava leaves [32]	Honeydew honeys [91]	Mung bean [81]	Olive oil [44]	Rice [47]	Teas [48]	Teff [55]
		Apple [60]	Blueberry [60]	Black chokeberry [60]	Black-berry [60]	Sour cherry [60]	Raspberry [60]	Strawberry [95]							
llic acid			26	2.2	96		47		142	0.32	54		1.7	5760	14
ydroxy-zoic acid		2.4	3.1	5.4	4.0	16	44			0.016					
lorogenic acid	1745	82	396	396		355				0.3	144			11355	
ffeic acid	2.6	1.6	47	30	3.1	3.2	2.0			0.3	200	1.2	13	760	2.5
ingic acid										0.33			11		2.5
tin	410			33				2.2		0.20			2	5795	105
Coumaric acid	50		14	1.5		1.6	3.0	1.0		0.39	233	0.6	20		113
apic acid			1.1	0.3	2.6	3.1	2.7							1870	
ulic acid	15									0.26	314		99	1360	160
speridin										0.10					
zoic acid										6.2					
icylic acid										1.4					
eolin										0.08					
ercetin	1450		11	53	10	45	11		390	0.35	82		20		27
nillic acid	73	2.5	6.4	5.6	5.4	11	1.3						23		7.2
ingenin		0.4	0.9	0.2	0.45	3.2	0.9		1078	0.04					
obanksin										0.05					
empferol						3.6		0.06		0.17	31				
hamnetin										0.10					
echin		2.5	3.1		2.0	26	3.5	0.9	7179		20		22		23
catechin		31	35	6.6	45	123	66							3140	
gallo-echin gallate									3.6					11040	
nic acid	2.6														
yl gallate	16														
tocatechuic d	0.25		36	45	21	22		0.16				2.1	41		102
namic acid													0.9		2.6
veratrol															3.7
gic acid			13	9	137		28	1.3	2116						

Overview of Electrochemical Methods

Electrochemical methods are based on measures of electrical parameters such as voltage, current, resistance, impedance, and charge, among others. The origin of these electrical signals comes from the conversion of the chemical signal by a transducer.

Conductivity measurement, which emerged in the 18th century, can be considered the oldest electroanalytical method, and just when this method was becoming widely applied, the procedure changed when Friedrich Kohlrausch incorporated the use of alternating current instead of direct current in the middle of that century [97]. In 1864, Wolcott Gibbs proposed electrogravimetry as a quantitative method for determining copper in coins containing copper and nickel [97, 98]. Coulometry, the measurement of electrical charges, was developed by Michael Faraday in 1834, wherein he constructed the first device for this proposed method [97, 99, 100]. Nowadays, conductometry, electrogravimetry, and coulometry are methods limited to just a few applications.

Currently, research in the electroanalytical area has been directed towards the development of potentiometric and amperometric sensors. Potentiometry and amperometry use a system containing, at least one working and one reference electrode. In amperometry, a third sensor called an auxiliary electrode was introduced to divert the flow of electrical current from the reference electrode to the auxiliary electrode in order to avoid secondary reactions in the latter [101]. The reference electrode is a device that maintains a constant potential during measurements done with the working electrode, wherein the chemical phenomenon of interest occurs at the latter. Nowadays, reference electrodes based on silver/silver chloride and 3 mol L^{-1} of potassium chloride solution are more common, because old reference electrodes were manufactured with thallium or mercury, chemicals that have been banned due to their toxicity [102].

Potentiometry appeared in 1889 with Nernst's work about the electromotive efficiency of ions [97, 103]. The start of the golden age for potentiometry occurred in the middle of the decade of the 1960´s with the development of ion selective electrodes based on a liquid membrane [104-

105]. After a period with few innovations between 1980 and 1990, potentiometry re-emerged with the invention of high-sensitivity solid-contact sensors that allowed for ion detection at sub-nanomolar levels [106]. Amperometry started with studies about current-voltage curves made by Ernst Salomon in 1896 [97, 107], and based on these studies the polarographic method was developed by Jaroslav Heyrovský who was awarded the Nobel Prize for chemistry in 1959 [97, 108]. In 1941, Laitinen and Kolthoff developed voltammetry, which is the part of the science of electrochemistry that aims to determine and interpret current-voltage curves [97, 109].

Biosensors are devices that are frequently applied in electroanalytical methods for the determination of phenolic compounds. These devices recognize a biological molecular substance through a transducer which is coupled to a biologically sensitive material that is called bioreceptor. Bioreceptors can be antigens, antibodies, DNA molecules, microorganisms, cells, organelles, tissues, enzymes and biomimetic materials. Biomimetic substances have the property of being able to imitate a biological compound, its working mechanisms, and other processes [110-112].

ELECTROCHEMICAL METHODS FOR PHENOL ANALYSIS

There are many electrochemical methods that have been developed for determining phenolic compounds, but few of these have been applied to food analysis. Ortega et al. developed an amperometric biosensor using tyrosinase covalently immobilized on the surface of an activated graphite electrode. Thirty phenolic compounds were analyzed using this sensor in a flow injection system. The detection limit for phenol was 3×10^{-9} mol L^{-1} with a quantification limit of 1×10^{-8} mol L^{-1}. Although this biosensor associated to flow injection system was not applied to real samples, the majority of the phenolic substances studied are present in foods [113]. Ortega and co-workers also presented the idea of using the biosensor as a detector in liquid chromatography for determining phenolic compounds in wastewater. The transducer for manufacturing of the biosensor was carbon paste and wastewater samples obtained from the pulp industry. The results showed the possibility of obtaining a more selective detection system of phenolic compounds *post*-column using the biosensor than with a diode array detector at 270 nm [114].

Another strategy for the preparation of sensitive amperometric biosensors for phenolic compounds was to immobilize the enzyme tyrosinase on a thin silica layer using the sol-gel technique. Carbon paste was used as a transducer and a homogeneous stock sol-gel solution was obtained through a mixture of tetramethoxysilane, cetyltrimethyl-ammonium bromide and methanol using alkaline catalysis. The limit of detection for catechol was 1.2×10^{-7} mol L^{-1} with this amperometric biosensor [115].

An amperometric biosensor was proposed by Sotomayor et al. using a biomimetic material to simulate the dopamine β-monooxygenase enzyme. The biomimetic substances, copper phtalocyanine and histidine (1:22), were mixed with graphite and mineral oil to produce a carbon paste. The biosensor presented a detection limit between 10 and 25 μmol L^{-1} for catechol, dopamine, guaiacol and serotonin [116].

The transducer applied to electrochemical measurements of phenolic compounds was also the subject of study of Kiralp et al. [56]. In this study, transducers made of conducting polymers instead of carbon paste were employed to immobilize polyphenol oxidase, and phenolic compounds were determined in wines through use of the biosensor. Another study used nanostructured materials (multiwall carbon, nanotube, fullerene and hydroxylated fullerene) as transducers for immobilization of the enzyme horseradish peroxidase in the development of amperometric biosensors for the determination phenolic compounds. These biosensors presented an operational stability for greater than 200 analyses, and the response range was between 5 and 200 μmol L^{-1} for phenol [117].

Screen-printed carbon sensors modified with graphene were manufactured to study the electrochemical behavior of gallic and ellagic acid using differential pulse voltammetry. These sensors have been employed to estimate the total content of phenolic compounds of low oxidation potential in cork boiling water. The data agreed with that found through Folin-Ciocalteau method, and the detection limit was lower than 0.1 ppm for both phenolic compounds [30].

Cyclic voltammetry was also used to quantify antioxidants in white and red wines. Carbon and silver/silver chloride electrodes were employed as working and reference electrodes, respectively. Wine samples presented a first peak close to an anode potential of 400 mV, and this peak was due to phenolic compounds containing an *o*-diphenolic group such as gallic acid. Kilmartin et al. observed a small peak at 300 mV in red wines due to the higher amount of myricetin, and quercetin glycosides showed one peak at 470 mV. Red and white wines differed by the presence of a peak at 640 mV in the red wine samples, and this peak was associated with malvidin and anthocyanidin molecules [118, 119].

Miniaturization of electrochemical systems has also been developed for determining phenol substances. Gonzales-Rivera and Osma manufactured an amperometric biosensor based on laccase for *in situ* determination of phenolic compounds by flow-Injection. The flow system was fabricated in reduced scale with 35mm of length and a width of 25 mm. The enzyme was immobilized on the working sensor by potential step chronoamperometry. The working electrode (200 μm Ø), reference electrode (300 μm Ø) and the counter electrode (ellipse format: 500/1000 μm Ø) consisted of gold electrodes. These electrodes were fabricated through photolithography, wherein a thin film of chrome and gold were deposited on glass slides by physical vapor deposition. The microfluidic system associated with the amperometric biosensor presented a detection limit of 1.5×10^{-7} mol L^{-1} for syringaldazine [120].

Besides amperometric sensors, potentiometric sensors are the only other ones that have been developed to measure phenolic compounds by electrochemical methods. A potentiometric sensor for dopamine was produced using a biomimetic material that consisted of copper salt entrapped in a polymeric membrane of ethylene-co-vinyl acetate. The working principle of this sensor was based on changing of the charge density on the sensor surface due to reduction of copper (II) to copper (I) by the dopamine. This sensor presented a super-Nernstian response, and the detection limit was 9×10^{-4} mol L^{-1} in conditions of flow analysis [121].

One potentiometric biosensor for determining adrenaline concentration was developed using the enzyme polyphenol oxidase extracted from banana peels and occluded in a carbon paste. This sensor presented a high sensitivity to adrenaline, about 8×10^{-9} mol L^{-1} [122].

The idea of using a natural source for obtaining enzymatic extracts of tyrosinase was also used for the development of a label-free potentiometric biosensor for determining phenolic compounds in honey and propolis [40]. This biosensor was manufactured according to a solid contact concept wherein the transducer has an intermediate layer with redox characteristics [106]. On this layer a composite of graphite and carboxylated poly(vinyl chloride) was deposited, with the enzyme tyrosinase being immobilized by a covalent bond. This potentiometric biosensor presented a low detection limit of 7.3×10^{-7} mol L^{-1} and a large linear range of response to catechol concentration from 9.3×10^{-7} to 8.3×10^{-2} mol L^{-1}. The measurements of the amount of total phenols in honey and propolis using this biosensor agreed with the Folin-Ciocalteau method [40].

CONCLUSION

Phenolic compounds have high antioxidant activity *in vitro* and *in vivo*. A diet rich in polyphenols has been correlated with a reduced risk of cancers and attacks by viruses. Perhaps for this reason, in modern, developed societies over the last few decades there has been an increase in the consumption of foods containing these substances. The conventional methods for determining phenolic compounds are still lacking in specific methodological protocols, are complex, and require a long analysis time. Hence, there is enormous potential in this field of research to develop new analytical methods to measure phenolic profiles in food.

Electrochemical biosensors are promising tools for phenol analysis since they are low-cost, highly sensitive, and enable efficient analysis. Although a separation method is needed to associate these devices for the determination of phenolic profiles in food, information about these phenolic profiles has increased in the last few years, and has allowed for the use of multivariate calibration methods with biosensors. Therefore, in the coming years it is expected that the development of commercial, disposable biosensors for verifying, for example, the adulteration of food based on its phenolic profile will become a promising field of research and development.

REFERENCES

[1] Harvey, D. *Modern Analytical Chemistry*, McGraw-Hill Higher Education: Boston, 2000, pp. 368.

[2] Kirchhoff, G. R.; Bunsen, R. W. Chemische Analyse durch Spectral beobachtungen. [Chemical analysis by spectral observations]. *Pogg. An.* 1860, *110*, 161-189.

[3] Miller, F. A. The history of spectroscopy as illustrated on stamps. *Appl. Spectrosc.* 1983, *37(3)*, 219-225.

[4] Koizumi, H.; Yasuda, K. An application of the Zeeman Effect to atomic absorption spectrometry: a new method for background correction. *Spectrochim. Acta B.* 1976, *31(5)*, 237-255.

[5] Bohr, N. On the Constitution of Atoms and Molecules, Part I. *Philos. Mag.* 1913, *26 (151)*, 1–24.

[6] Raman, C. V.; Krishnan, K. S. A New Type of Secondary Radiation. *Nature* 1928, *121*, 501-502.

[7] Simoni, R. D.; Hill, R. L.; Vaughan, M.; Tabor, H. A Classic Instrument: The Beckman DU Spectrophotometer and its Inventor, Arnold O. Beckman. *J. Biol. Chem.* 2003, *278(49)*, 79-81.

[8] Beckman, A. O.; Gallaway, W. S.; Kaye, W.; Ulrich, W. F. History of Spectrophotometry at Beckman Instruments, Inc. *Anal. Chem.* 1977, *49(3)*, 280A–300A.

[9] Folin, O.; Ciocalteau, V. On tyrosine and tryptophane determinations in proteins. *J. Biol. Chem.* 1927, *73*, 627-650.

[10] Folin, O.; Denis, W. On phosphotungstic-phosphomolybdic compounds as colour reagents. *J. Biol. Chem.* 1912, *12*, 239-243.

[11] Folin, O.; Looney, J. M. Colorimetric methods for the separate determination of tyrosine, tryptophane, and cystine in proteins. *J. Biol. Chem.* 1922, *51*, 421-434.

[12] Box, J. D. Investigation of the Folin-Ciocalteau phenol reagent for the determination of polyphenolic substances in natural waters. *Water Res.* 1983, *17(5)*, 511-525.

[13] Sousa, C. M. M.; Silva, H. R.; Vieira-Jr, G. M.; Ayres, M. C. C.; Costa, C. L. S.; Araújo, D. S.; Calvalcante, L. C. D.; Barros, E. D. S.; Araújo, P. B. M.; Brandão, M. S.; Chaves, M. H. Fenóis totais e atividade antioxidante de cinco plantas medicinais. [Total phenolics and antioxidant activity of five medicinal plants]. *Quim. Nova* 2007, *30(2)*, 351-355.

[14] Ikawa, M.; Schaper, T. D.; Dollard, C. A.; Sasner, J. J. Utilization of Folin-Ciocalteu phenol reagent for the detection of certain nitrogen compounds. *J. Agric. Food Chem.* 2003, *51(7)*, 1811-1815.

[15] Naczk, M.; Shahidi, F. Extraction and analysis of phenolics in food. *J. Chromatogr. A* 2004, *1054(1-2)*, 95-111.

[16] Bonoli, M.; Verardo, V.; Marconi, E.; Caboni, M. F. Antioxidant phenols in barley (*Hordeum vulgare L.*) flour: Comparative spectrophotometric study among extraction methods of free and bound phenolic compounds. *J. Agric. Food Chem.* 2004, *52*, 5195-5120.

[17] Everette, J. D.; Bryant, Q. M.; Green, A. M.; Abbey, Y. A.; Wangila, G. W.; Walker, R. B. A thorough study of reactivity of various compound classes towards the Folin-Ciocalteu reagent. *J. Agric. Food Chem.* 2010, 58(14), 8139–8144.

[18] Barrows, J. N.; Jameson, G. B.; Pope, M. T. Structure of a heteropoly blue. The four-electron reduced β-12-molybdophosphate anion. *J. Am. Chem. Soc.* 1985, *107*, 1771-1773.

[19] Wikipedia contributors, "Polyoxometalate", Wikipedia, The Free Encyclopedia, https://en.wikipedia.org/w/index.php?title=Polyoxometalate&oldid =766562116 (accessed March 10, 2017).

[20] Müller, A.; Serain, C. Soluble Molybdenum Blues -"des Pudels Kern". *Acc. Chem. Res.* 2000, *33*, 2-10.

[21] Hall, N. Bringing inorganic chemistry to life. *Chem. Commun.* 2003, 803-806.

[22] Kloster, M. B. The determination of tannin and lignin. *J. Am. Wat. Wks. Ass.* 1974, *66*, 44-46.

[23] Batista, C. C. R.; de Oliveira, M. S.; Araújo, M. E.; Rodrigues, A. M. C.; Botelho, J. R. S., Filho, A. P. S. S.; Machado, N. T.; Carvalho-Junior, R. N. Supercritical CO2 extraction of açaí (*Euterpe oleracea*) berry oil: Global yield, fatty acids, allelopathic activities, and determination of phenolic and anthocyanins total compounds in the residual pulp. *J. Supercrit. Fluids* 2016, *107*, 364-369.

[24] Wang, X.; Li, C.; Liang, D.; Zou, Y.; Li, P.; Ma, F. Phenolic compounds and antioxidant activity in red-fleshed apples. *J. Funct. Foods* 2015, *18*, 1086-1094.

[25] Gómez, F. S.; Sánchez, S. P.; Iradi, M. G. G.; Azman, N. A. M.; Almajano, M. P. Avocado Seeds: Extraction Optimization and Possible Use as Antioxidant in Food. *Antioxidants* 2014, *3*, 439-454. doi:10.3390/antiox3020439.

[26] Segovia, F. J.; Corral-Pérez, J. J. Almajano, M. P. Avocado seed: Modeling extraction of bioactive compounds. *Ind. Crop. Prod.* 2016, *85*, 213–220.

[27] Vinson, J. A.; Mandarano, M.; Hirst, M.; Trevithick, J. R.; Bose, P. Phenol antioxidant quantity and quality in foods: Beers and the effect of two types of beer on an animal model of atherosclerosis. *J. Agric. Food Chem.* 2003, *51*, 5528-5533.

[28] Cascant, M. M.; Sisouanen, M.; Tahiri, S.; Krati M. E.; Cervera, M. L.; Garrigues, S.; de la Guardia, M. Determination of total phenolic compounds in compost by infrared spectroscopy. *Talanta* 2016, *153*, 360-365.

[29] Hu, Y.; Pan, Z. J.; Liao, W.; Li, J.; Gruget, P.; Kitts, D. D.; Lu, X. Determination of antioxidant capacity and phenolic content of chocolate by attenuated total reflectance-Fourier transformed-infrared spectroscopy. *Food Chem.* 2016, *202*, 254-261.

[30] Guiberteau-Cabanillas, A.; Godoy-Cancho, B.; Bernalte, E.; Tena-Villares, M.; Guiberteau-Cabanillas, C.; Martínez-Cañas, M. A. Electroanalytical behavior of gallic and ellagic acid using graphene modified screen-printed electrodes. Method for the determination of total low oxidation potential phenolic compounds content in cork boiling waters. *Electroanal.* 2015, *27*, 177-184.

[31] Amaya-Cruz DM., Rodríguez-González S, Pérez-Ramírez IF, Loarca-Piña G, Amaya-Llano S, Gallegos-Corona MA, Reynoso-Camacho R. Juice by-products as a source of dietary fibre and antioxidants and their effect on hepatic steatosis. *J. Funct. Foods* 2015, *17*, 93-102.

[32] Díaz-de-Cerio, E.; Gómez-Caravaca, A. M.; Verardo, V.; Fernández-Gutiérrez, A.; Segura-Carretero, A. Determination of guava (*Psidium guajava L.*) leaf phenolic compounds using HPLC-DAD-QTOF-MS. *J. Funct. Foods* 2016, *22*, 376-388.

[33] Chen, H-Y.; Yen, G-C. Antioxidant activity and free radical-scavenging capacity of extracts from guava (Psidium guajava L.) leaves. *Food Chem.* 2007, *101(2)*, 686–694.

[34] da Silva, L. M. R.; de Figueiredo, E. A. T.; Ricardo, N. M. P. S.; Vieira, I. G. P.; de Figueiredo, R. W.;, Brasil, I. M.; Gomes, C. L. Quantification of bioactive compounds in pulps and by-products of tropical fruits from Brazil. *Food Chem.* 2014, *143*, 398-404.

[35] Thuaytong, W. Anprung, P. Bioactive compounds and prebiotic activity in Thailand-grown red and white guava fruit (*Psidium guajava L.*). *Food Sci. Technol. Int.* 2011, *17(3)*, 205–212.

[36] Thaipong, K.; Boonprakob, U.; Crosby, K.; Cisneros-Zevallos, L.; Byrne, D. H. Comparison of ABTS, DPPH, FRAP, and ORAC assays for estimating antioxidant activity from guava fruit extracts. *J. Food Comp. Anal.* 2006, *19(6–7)*, 669-675.

[37] Luximon-Ramma, A.; Bahorun, T.; Crozier, A. Antioxidant actions and phenolic and vitamin C contents of common Mauritian exotic fruits. *J. Sci. Food Agr.* 2003, *83(5)*, 496-502.

[38] Haida, K. S.; Baron, Â.; Haida, K. S.; de Faci, D.; Haas, J. Phenolic compounds and antioxidant activity of two varieties of guava and rue. *Reva. Bra. Ciênc. Saúde* 2011, *28*, 11-19.

[39] Chis, A. M.; Purcarea, C. Versatility of Folin-Ciocalteu method applied on honey determination of total phenols. *Rev. Chim.* 2016, *67(10)*, 1932-1935.

[40] Draghi, P. F.; Fernandes, J. C. B. Label-free potentiometric biosensor based on solid-contact for determination of total phenols in honey and propolis. *Talanta* 2017, *164*, 413-417.

[41] López-Cobo, A.; Gómez-Caravaca, A. M.; Švarc-Gajić, J.; Segura-Carretero, A.; Fernández-Gutiérrez, A. Determination of phenolic

compounds and antioxidant activity of a Mediterranean plant: The case of *Satureja montana subsp. Kitaibelii*. *J. Funct. Foods* 2015, *18*, 1167–1178.

[42] Gan, R-Y.; Wang, M-F.; Lui, W-Y.; Wu, K.; Corke, H. Dynamic changes in phytochemical composition and antioxidant capacity in green and black mung bean (*Vigna radiata*) sprouts. *Int. J. Food Sci. Technol.* 2016, *51*, 2090-2098.

[43] Lara-Ortega, F. J.;.Sainz-Gonzalo, F. J.; Gilbert-López, B.; García-Reyes, J. F.; Molina-Díaz, A. Multicommuted flow injection method for fast photometric determination of phenolic compounds in commercial virgin olive oil samples. *Talanta* 2016, *147*, 531-536.

[44] del Monaco, G.; Officioso, A.; D'Angelo, S.; La Cara, F.; Ionata, E.; Marcolongo, L.; Squillaci, G.; Maurelli, L.; Morana, A. Characterization of extra virgin olive oils produced with typical Italian varieties by their phenolic profile. *Food Chem.* 2015, *184*, 220-228.

[45] Antonini, E.; Farina, A.; Leone, A.; Mazzara, E.; Urbani, S.; Selvaggini, R.; Servili, M.; Ninfali, P. Phenolic compounds and quality parameters of family farming versus protected designation of origin (PDO) extra-virgin olive oils. *J. Food Compost. Anal.* 2015, *43*, 75-81.

[46] González, M.; Guzmán, B.; Rudyk, R.; Romano, E.; Molina, M. A. A. Spectrophotometric determination of phenolic compounds in propolis. *Lat. Am. J. Pharm.* 2003, *22(3)*, 243-248.

[47] Sumczynski D, Kotásková E, Družbíková H, Miček J. Determination of contents and antioxidant activity of free and bound phenolics compounds and in vitro digestibility of commercial black and red rice (*Oryza sativa L.*) varieties. *Food Chem.* 2016, *211*, 339-346.

[48] Bae IK, Hamb HM, Jeong MH, Kim DH, Kim HJ. Simultaneous determination of 15 phenolic compounds and caffeine in teas and mate using RP-HPLC/UV detection: Method development and optimization of extraction process. *Food Chem.* 2015, *172*, 469-475.

[49] Jeszka-Skowron, M.; Krawczyk, M.; Zgoła-Grześkowiak, A. Determination of antioxidant activity, rutin, quercetin, phenolic acids and trace elements in tea infusions: Influence of citric acid addition on extraction of metals. *J. Food Compost. Anal.* 2015, *40*, 70–77.

[50] Oh, J.; Jo, H.; Cho, A. R.; Kim, S-J.; Han, J. Antioxidant and antimicrobial activities of various leafy herbal teas. *Food Control.* 2013, *31*, 403-409.

[51] Bae, I. K.; Hamb, H. M.; Jeong, M. H.; Kim, D. H.; Kim, H. J. Simultaneous determination of 15 phenolic compounds and caffeine in teas and mate using RP-HPLC/UV detection: Method development and optimization of extraction process. *Food Chem.* 2015, *172*, 469-475.

[52] Frizon, C. N. T.; Oliveira, G. A.; Perussello, C. A.; Peralta-Zamora, P. G.; Camlofski, A. M. O.; Rossa, Ü. B.; Hoffmann-Ribani, R. Determination of total phenolic compounds in yerba mate (*Ilex paraguariensis*) combining near infrared spectroscopy (NIR) and multivariate analysis. *LWT - Food Sci. Technol.* 2015, *60*, 795-801.

[53] Bhebhe, M.; Chipurura, B.; Muchuweti, M. Determination and comparison of phenolic compound content and antioxidant activity of selected local Zimbabwean herbal teas with exotic *Aspalathus linearis*. *S. Afr. J. Bot.* 2015, *100*, 213-218.

[54] Ariffin, F.; Chew, S. H.; Bhupinder, K.; Karim, A. A.; Huda, N. Antioxidant capacity and phenolic composition of fermented *Centella asiatica* herbal teas. *J. Sci. Food Agric.* 2011, *91*, 2731-2739.

[55] Kotásková, E.; Sumczynski, D.; Miček, J.; Valášek, P. Determination of free and bound phenolics using HPLC-DAD, antioxidant activity and *in vitro* digestibility of *Eragrostis tef*. *J. Food Compost. Anal.* 2016, *46*, 15-21.

[56] Kıralp, S.; Toppare, L.; Yağcı, Y. Immobilization of polyphenol oxidase in conducting copolymers and determination of phenolic compounds in wines with enzyme electrodes. *Int. J. Biol. Macromolec.* 2003, *33*, 37-41.

[57] Karakaya, S.; El, S. N.; Tas, A. A. Antioxidant activity of some foods containing phenolic compounds. *Int. J. Food Sci. Nutr.* 2001, *52(6)*, 501-508.

[58] Olejar, K. J.; Fedrizzi, B.; Kilmartin, P. A. Enhancement of Chardonnay antioxidant activity and sensory perception through maceration technique. *LWT - Food Sci. Technol.* 2016, *65*, 152-157.

[59] Sandoval-Ventura, O.; Olguín-Contreras, L. F.; Cañizares-Macías, M. P. Total polyphenols content in white wines on a microfluidic

flow injection analyzer with embedded optical fibers. *Food Chem.* 2017, *221*, 1062-1068.

[60] Čakar, U. D.; Petrović, A. V.; Živković, M. B.; Vajs, V. E.; Milovanović, M. M.; Zeravik, J.; Đorđević, B. I. Phenolic profile of some fruit wines and their antioxidant properties. *Hem. Ind.* 2016, *70(6)*, 661–672.

[61] Del Pino-García, R.; González-SanJosé, M. L.; Rivero-Pérez, M. D.; García-Lomillo, J.; Muñiz, P. The effects of heat treatment on the phenolic composition and antioxidant capacity of red wine pomace seasonings. *Food Chem.* 2017, *221*, 1723-1732.

[62] Singer, A. J.; Stern, E. R.; Spectrophotometric Determination of Bischlorophenol and other Phenolic Compounds. *Anal. Chem.* 1951, *23(10)*, 1511-1512.

[63] Gottlieb, S.; Marsh, P. B.; Quantitative determination of phenolic fungicides. *Ind. Eng. Chem. Anal. Ed.*, 1946, *18(1)*, 16-19.

[64] Gibbs, H. D. Phenol tests. III. The indophenol test. *J. Biol. Chem.* 1927, *72*, 649-664.

[65] Večera, M.; Gasparič, J. *Detection and Identification of Organic Compounds.* Plenum Press: New York, 1971, pp. 196.

[66] Kolliopoulos, A. V.; Kampouris, D. K.; Banks. C. E. Indirect electroanalytical detection of phenols. Analyst 2015, *140*, 3244-3250.

[67] Mohler-Jr., E. F.; Jacob L. N. Determination of Phenolic-Type Compounds in Water and Industrial Waste Waters: Comparison of Analytical Methods. *Anal. Chem.* 1957, *29(9)*, 1369-1374.

[68] Fountaine, J. E.; Joshipura, P. B.; Keliher P. N.; Johnson, J. D. New Ultraviolet Ratio Spectrophotometric System for the Determination of Trace Amounts of Phenolic Compounds. *Anal. Chem.* 1974, *46(1)*, 62-66.

[69] Frenzel, W; Frenzel, J. O.; Moller, J. Spectrophotometric determination of phenolic compounds by flow-injection analysis. *Anal. Chim. Acta* 1992, *261*, 253-259.

[70] Lupetti, K. O.; Rocha F. R. P.; Fatibello-Filho O. An improved flow system for phenols determination exploiting multicommutation and long pathlength spectrophotometry. *Talanta* 2004, *62*, 463–467.

[71] Martens, H.; Naes, T. *Multivariate Calibration*. John Wiley & Sons: New York, 1992, 419 p.

[72] Jørgensen, B.; Goegebeur, Y. [http://statmaster.sdu.dk/courses/ST02, *Module from 2 to 8*, January 29, 2007], (accessed February 15, 2015).

[73] Manne, R. Analysis of two partial-least-squares algorithms for multivariate calibration. *Chemometr. Intell. Lab*. 1987, *2*, 187-197.

[74] Wold, S.; Sjostrom, M.; Eriksson, L. PLS-regression: A Basic tool of chemometrics. *Chemom Intell. Lab.* 2001, *58*, 109–130.

[75] Geladi, P.; Kowalski, R. B. Partial Least Squares regression: A Tutorial. *Anal. Chim. Acta* 1986, *185*, 1-17.

[76] Beebe, K.; Kowalski, B. R. An Introduction to Multivariate Calibration and Analysis. *Anal. Chem.* 1987, *59(17)*, 1007A-1017A.

[77] Brereton, R. G. *Applied Chemometrics for Scientists*. John Wiley & Sons: Chichester, 2007, pp. 211-214.

[78] Jong, S. T. SIMPLS: An alternative approach to partial least squares regression. *Chemom. Intell. Lab.* 1993, *18*, 251-263.

[79] Martelo-Vidal, M. J.; Vázquez, M. Determination of polyphenolic compounds of red wines by UV–VIS–NIR spectroscopy and chemometrics tools. *Food Chem.* 2014, *158*, 28-34.

[80] Mazur, L., Peralta-Zamora P. G., Demczuk-Jr., B., Ribani, R. H. Application of multivariate calibration and NIR spectroscopy for the quantification of methylxanthines in yerba mate (*Ilex paraguariensis*). *J. Food Comp. Anal.* 2014, *35*, 55-60.

[81] Meenu, M.; Kamboj, U.; Sharma, A.; Guha, P.; Mishra, S. Green method for determination of phenolic compounds in mung bean (*Vigna radiata L.*) based on near-infrared spectroscopy and chemometrics. *Int. J. Food Sci. Tech.* 2016, *51*, 2520-2527.

[82] Tswett, M. Physikalisch-chemische Studien über das Chlorophyll. Die Adsorptionen. [Physical-chemical studies on chlorophyll. The adsorption]. *Ber. Dtsch. Bot. Ges.* 1906, *24*, 316-323.

[83] Cirillo V. J. The history of gas-Liquid chromatography. *J. Chromatography* 1973, *81*, 197-205.

[84] Martin, A. J. P.; Synge, R. L. M. A new form of chromatogram employing two liquid phases. 1. A theory of chromatography. 2. Application to the micro-determination of the higher monoamino-acids in proteins. *Biochem. J.* 1941, *35*, 1358-1368.

[85] James, A. T.; Martin, A. J. P. Gas-liquid partition chromatography: The separation and micro-estimation of volatile fatty acids from formic acid to. dodecanoic Acid. *Biochem. J.* 1952, *50*, 679-690.

[86] Klesper, K.; Corwin, A. H.; Turner, D. A. High pressure gas chromatography above critical temperatures. *J. Org. Chem.* 1962, *27*, 700-701.

[87] Saito, M. History of supercritical fluid chromatography: Instrumental development. *J. Biosci. Bioeng.* 2013, *115(6)*, 590-599.

[88] Karger, B. L. HPLC: Early and Recent Perspectives. *J. Chem. Educ.* 1997, 74, 45-48.

[89] Dolatto R. G.; Messerschmidt I.; Pereira B. F.; Martinazzo R.; Abate G. Preconcentration of polar phenolic compounds from water samples and soil extract by liquid-phase micro extraction and determination via liquid chromatography with ultraviolet detection. *Talanta* 2016, *148*, 292-300.

[90] López-Cobo, A.; Gómez-Caravaca, A. M.; Pasini, F.; Caboni, M. F.; Segura-Carretero, A.; Fernández-Gutiérrez, A. HPLC-DAD-ESI-QTOF-MS and HPLC-FLD-MS as valuable tools for the determination of phenolic and other polar compounds in the edible part and by-products of avocado. *LWT - Food Sci. Technol.* 2016, *73*, 505-513.

[91] Seraglio, S. K. T.; Valese, A. C.; Daguer H.; Bergamo, G.; Azevedo, M. S.; Gonzaga, L. V.; Fett, R.; Costa, A. C. O. Development and validation of a LC-ESI-MS/MS method for the determination of phenolic compounds in honeydew honeys with the diluted-and-shoot approach. *Food Res. Int.* 2016, *87*, 60-67.

[92] Wua, R.; Ma, F.; Zhang, L.; Li, P.; Li, G.; Zhang, Q.; Zhang, W.; Wanga, X. Simultaneous determination of phenolic compounds in sesame oil using LC–MS/MS combined with magnetic carboxylated multi-walled carbon nanotubes. *Food Chem.* 2016, *204*, 334-342.

[93] Kang, J.; Price, W. E.; Ashton, J.; Tapsell, L. C.; Johnson, S. Identification and characterization of phenolic compounds in hydromethanolic extracts of sorghum wholegrains by LC-ESI-MS^n. *Food Chem.* 2016, *211*, 215-226.

[94] Van der Kamp, J. W.; Jones, J.; McCleary, B.; Topping, D. *Dietary fibre: New frontiers for food and health*, Wageningen Academic Publishers: The Netherlands, 2010, pp. 422.

[95] Álvarez-Fernández, M. A.; Hornedo-Ortega, R.; Cerezo, A. B.; Troncoso, A. M.; García-Parrilla, M. C. Determination of nonanthocyanin phenolic compounds using high-resolution mass spectrometry (UHPLC-Orbitrap-MS/MS) and impact of storage conditions in a beverage made from strawberry by fermentation. *J. Agric. Food Chem.* 2016, *64*, 1367-1376.
[96] Figueiredo-González, M.; Cancho-Grande, B.; Simal-Gándara, J.; Teixeira, N.; Mateus, N.; De Freitas, V. The phenolic chemistry and spectrochemistry of red sweet wine-making and oak-aging. *Food Chem.* 2014, *152*, 522-530.
[97] Lubert, K-H.; Kalcherb, K. History of Electroanalytical Methods. *Electroanal.* 2010, *22(17-18)*, 1937-1946.
[98] Gibbs, W. Beiträge zur Chemie aus dem Laboratorium der Lawrence Scientific School: VI. Ueber die Anwendung der elektrolytischen Ausfällung von Kupfer und Nickel in der Analyse. [Contributions to chemistry from the laboratory of the Lawrence Scientific School: VI. On the application of the electrolytic precipitation of copper and nickel in the analysis]. *Fresenius Z. Anal. Chem.* 1864, *3*, 334-336.
[99] Faraday, M. Experimental researches in electricity – Sixth series. *Phil. Trans. R. Soc. Lond.* 1834, 124, 55-76.
[100] Faraday, M. Experimental researches in electricity – Seventh series. *Phil. Trans. R. Soc. Lond.* 1834, 124, 77-122.
[101] Harris, D. C. *Quantitative chemical analysis*, W. H. Freeman and Company: New York, 2007, pp. 353.
[102] Fernandes, J. C. B.; Heinke, E. Alternative strategy for manufacturing of all-solid-state reference electrodes for potentiometry. *J. Sens. Sens. Syst.* 2015, *4*, 53–61. doi:10.5194/jsss-4-53-2015.
[103] Nernst, W. Die elektromotorische Wirksamkeit der Ionen. [The electromotive efficiency of the ions]. *Fresenius Z. Phys. Chem.* 1889, *4*, 129-181.
[104] Fernandes, J. C. B.; Kubota, L. T.; Neto, G. O. Eletrodos íon-seletivos: Histórico, mecanismo de resposta, seletividade e revisão dos conceitos. [Ion-selective electrodes: Historical, mechanism of response, selectivity and concept review]. *Quim. Nova* 2001, *24(1)*, 120-130.

[105] Frant, M. S.; Ross-Jr, J. W. *Ion-sensitive electrode with organometallic liquid ion-exchanger.* US 3,406,102, October 15, 1968.
[106] Bakker E.; Pretsch E. Modern Potentiometry. *Angew Chem. Int. Ed. Engl.* 2007, *46(30)*, 5660-5668.
[107] Salomon, E. Theorie des Reststromes, den man bei polarisierten Elektroden beobachtet. [Theory of the residual current observed with polarized electrodes]. *Z. Elektrochem.* 1896, *3(12)*, 264-266.
[108] Heyrovský, J. Elektrolysa se rtutovou kapkovou kathodou. [Electrolysis with a Dropping Mercury Cathode]. *Chem. Listy* 1922, *16*, 256-264.
[109] Laitinen, H. A.; Kolthoff, I. M. Voltammetry with stationary microelectrodes of platinum wire. *J. Phys. Chem.* 1941, *45*, 1061-1079.
[110] Cooper, M. A. Label-Free Biosensors: Techniques and Applications. 1st ed., Cambridge University Press: New York, 2009, 300p.
[111] Sadana, A.; Sadana, N. Handbook of Biosensors and Biosensor Kinetics. 1st ed., Elsevier: Amsterdam, 2011, 524p.
[112] Stroble, J. K.; Stone, R. B.; Watkins, S. E. An overview of biomimetic sensor technology. *Sensor Rev.* 2009, *29(2)*, 112-119. doi: 10.1108/02602280910936219.
[113] Ortega, F.; Domlnguez, E.; Jönsson-Pettersson, G.; Gorton, L. Amperometric biosensor for the determination of phenolic compounds using a tyrosinase graphite electrode in a flow injection system. *J. Biotechnol.* 1993, *31*, 289-300.
[114] Ortega, F.; Domlnguez, E.; Burestedt, E.; G.; Emnéus, J.; Gorton, L.; Marko-Varga, G. Phenol oxidase-based biosensors as selective detection units in column liquid chromatography for the determination of phenolic compounds. *J. Chromatog. A* 1994, *675*, 65-78.
[115] Li, J.; Chia, L. S.; Goh, N. K.; Tan, S. N. Silica sol-gel immobilized amperometric biosensor for the determination of phenolic compounds. *Anal. Chim. Acta* 1998, *362*, 203-211.
[116] Sotomayor, M. D. P. T.; Tanaka, A. A.; Kubota, L. T. Development of an enzymeless biosensor for the determination of phenolic compounds. *Anal. Chim. Acta* 2002, *455*, 215-223.

[117] Santos, A. S.; Costa, V. C.; Felício, R. C. Comparative study of nanostructured matrices employed in the development of biosensors based on HRP enzyme for determination of phenolic compounds. *Electroanal.* 2015, *27*, 1572-1578.

[118] Kilmartin, P. A.; Zou, H. L.; Waterhouse, A. L. Correlation of wine phenolic composition versus cyclic voltammetry response. *Am. J. Enol. Vitic.* 2002, *53*, 294-302.

[119] Dobes, J.; Zitka, O.; Sochor, J.; Ruttkay-Nedecky, B.; Babula, P.; Beklova, M.; Kynicky, J.; Hubalek, J.; Klejdus, B.; Kizek, R.; Adam, V. Electrochemical tools for determination of phenolic compounds in plants. A Review. *Int. J. Electrochem. Sci.* 2013, 8, 4520-4542.

[120] Gonzalez-Rivera, J. C.; Osma, J. F. Fabrication of an Amperometric Flow-Injection microfluidic biosensor based on laccase for in situ determination of phenolic compounds. *Biomed. Res. Int.* 2015, *Article ID 845261*, 9 p. doi:10.1155/2015/845261.

[121] Rover-Jr, L.; Fernandes, J. C. B.; Neto, G. O.; Kubota, L. T. Development of a new FIA-potentiometric sensor for dopamine based on EVA-copper(II) ions. *J. Electroanal. Chem.* 2000, *481*, 34-41.

[122] Mataveli, L. R. V.; Antunes, N. J.; Brigagão, M. R. P. L.; Magalhães, C. S.; Wisniewski, C.; Luccas, P. O. Evaluation of a simple and low cost potentiometric biosensor for pharmaceutical and *in vivo* adrenaline determination. *Biosens. Bioelectron.* 2010, *26*, 798-802.

Reviewed by:

Dr. Deyanira Fuentes Silva and Dr. Troy Patrick Beldini.

In: Phenolic Compounds
Editors: P. Oliver and A. Villem
ISBN: 978-1-53612-881-9

Chapter 2

THE ROLE OF BERRY GROWING TECHNOLOGY ON BIOACTIVE COMPOUND IMPROVEMENT

Jelena Tomić*[*], *Žaklina Karaklajić Stajić* and *Marijana Pešaković
Fruit Research Institut, Čačak, Republic of Serbia

ABSTRACT

Numerous studies have pointed out the presence of high bioactive substance concentration in fruit berries as well as a correlation between daily intake of fruits and human health. The greatest benefit to human health has been attributed to phenolic compounds and vitamin C, due to their antioxidant, anticarcinogenic, antimutagenic, antimicrobial, anti-inflammatory and neuroprotective properties. A significant part of antioxidant activity in strawberry and blackberry fruits takes place due to

[*] Corresponding Author E-mail: jtomic@institut-cacak.org.

phenolic acids. Apart from phenolic acids, flavonoids also make an important group of secondary metabolites, among which the most common in strawberries and blackberries are anthocyanins.

Many factors affect the chemical composition of berries, the most important of them being the genotype, climatic conditions, agricultural practices, fruit ripeness degree and others. The main objective of this study was to investigate the influence of genotype and type of fertilizer (strawberry), as well as the system of cultivation and climatic conditions (blackberry) on the content of vitamin C and phenolic compound.

The application of biofertilizers predominantly expressed a stimulating influence on most of the phenolic compounds in fruit berries tested. In the rain-shield cultivation of blackberries, higher values for all parameters of the fruit phenolic composition were obtained except for vanillic acid content, in comparison to standard cultivation system.

Changes that have occurred in the nutritional quality of the tested strawberry cultivars indicate the best performance was exhibited by 'Joly', which can be recommended for further promotion and expansion in strawberry growing regions. Taking into consideration the stimulating impact of biofertilizers in strawberries and rain-shield cultivation systems in blackberries on the phenolic compound in fruits, the use of biofertilization and rain-shield can be considered justified in terms of improving the current strawberry and blackberry production technology and obtaining products with positive effects on human health.

Keywords: strawberry, blackberry, cultivar, biofertilizer, rain shield, phenols

Antioxidant Activity of Bioactive Compounds in Berries

Berry fruits possess an outstanding nutritive potential based on the high content of antioxidant components (Ding et al., 2006; Tulipani et al., 2008). Antioxidants include vitamins, phenols, carotenoids, glutathione, and endogenous metabolites (Larson, 1988).

Numerous quotes that highlight the antioxidant activity of vitamin C can be found in literature (Omaye & Zhang, 1998; Szajdek & Borowska, 2008), mainly expressed synergetic relation with flavonoids (Isler et al., 1988; Kähkönen et al., 2001). Ascorbic acid easily releases electrons, reducing thus reactive oxygen compounds or free radicals (Klein & Kurilich, 2000; Prior & Cao, 2000). As a natural antioxidant, ascorbic acid is added to different foods for preventing or avoiding obscurity, discolouring, prolonging the storage life accordingly (Castro, 2009). Its influence is reflected in the prevention of darkening and discoloration of fruits, as well as increasing their lifetime (Voća et al., 2006), with the inhibitory effect on physiological processes that diminish the quality of the fruit during storage (Lattanzio et al., 2001). On the other hand, vitamin C affects the so-called reduction of oxidative stress in human body, thus reducing the risk of chronic diseases and cancer (Liu, 2003). According to Wang et al. (1997), participation of vitamin C in fruit products antioxidative capacity is approximately 15%.

In addition to vitamin C, berry fruits contain a large amount of phenolic compounds. Phenolic compounds are secondary metabolites of plants, present in a large number of plant species. Considering that about 8000 of them are known, they constitute one of the largest compound group in nature (Spanos & Wrolstad, 1992; Harborne & Baxter, 1999; Haminiuk et al., 2012). Their basic structure consists of a benzene ring to which one or more hydroxyl groups can be attached (Bravo, 1998). According to the number of aromatic nuclei in a molecule, there are: monophenols, which contain one benzene ring, with one or more hydroxyl groups attached to it (free phenols, phenolic acids and derivatives) and polyphenols, which contain a larger number of benzene rings in one molecule (flavonoids) (Macheix et al., 1990; Shahidi & Naczk, 2003; Maestri et al., 2006; Pereira et al., 2009; Katalinić et al., 2010). Due to the specific structure of a molecule, phenolic compounds have a strong potential for interaction with proteins, because of which some enzymes can be inhibited (lipoxygenase, cyclooxygenase, xanthine oxidase, etc.) and act as antioxidants (Cos et al., 1988; Parr & Bolwell, 2002).

Sumbul et al. (2011) report that phenols, according to their chemical composition, are comprised of non-flavonoids and flavonoids (Figure 1). The non-flavanods group includes hydroxybenzoic, hydroxycinnamic acids and stibenes. Hydroxybenzoic and hydroxycinnamic acids are different in the degree of hydroxylation and the methylation of aromatic ring (Macheix et al., 1990; Robbins, 2003). The group hydroxybenzene acids include gallic, *p*-hydroxybenzoic, vanillic, syringic, protocatehine, salicylic and ellagic acid (Macheix et al., 1990; Pereira et al., 2009). Hydroxycinnamic acids are rarely found in their free forms but mostly in conjugated forms, such as esters of *p*-coumaric acid, caffeic, ferulic and sinapic acids (Vasco, 2009). In the plant world the most common is caffeic acid (Pereira et al., 2009; Dai et al., 2010).

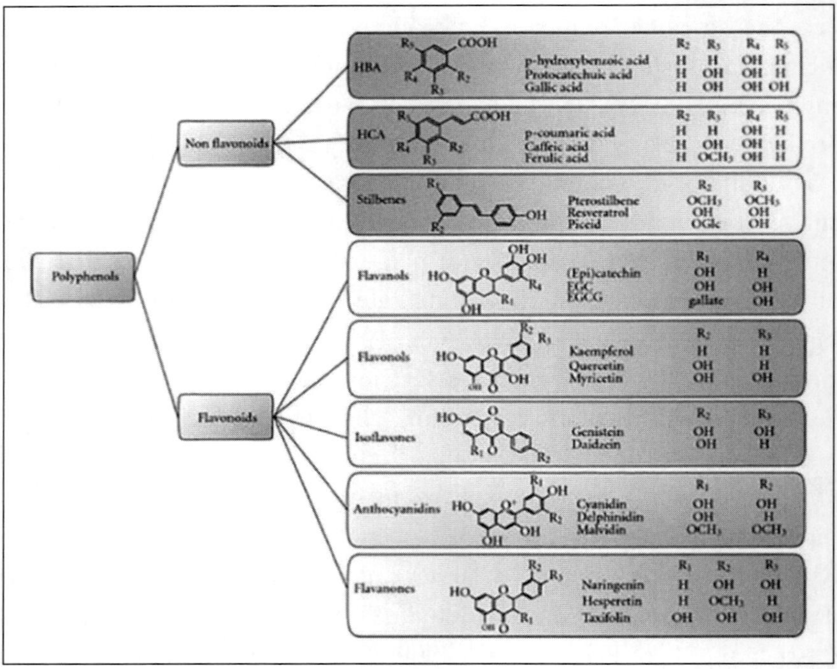

Figure 1. Structural hierarchy of phenolic compounds (Vauzour, 2012).

Flavonoids are the largest group of phenolic compounds (around 6400 identified), represented in a variety of plant species (Harborne & Baxter, 1999; Prochzáková et al., 2011). Flavonoids can be divided into 5 groups: flavonols, flavonols, isoflavones, anthocyanidins and flavanones.

Kostanesku in 1895 isolated the parent compound and named it "flavone" (Latin flavus-yellow), according to which a large group of chemically similar compounds are named flavonols. Flavonols have a common molecular structure, consisting of the tricyclic C6-C3-C6 'flavone skeleton'. They occur in the form of free or glycosidically linked cell yellow pigments localized in various organs of plants, particularly in fruits, flowers and leaves, distributed in plants usually as *O*-glycosides. Robards et al. (1999) report that there are over 200 flavonol aglycones identified in plants but only 4 characteristic of the fruit, and those are quercetin, kaempferol, myricetin and izorahmetin. Caridi et al. (2007) suggest that flavonols are accumulated in surface tissues of fruit plants, and therefore their contents, significantly influenced by the light regime and the conditions during storage and processing.

Anthocyanins are a subgroup of flavonoids, which are widely present in fruits and vegetables. Zhao (2007) points out that the anthocyanin is rarely present in nature as a free compound (anthocyanidins) due to its high volatility, therefore their occurrence in fruits is mostly in conjunction with different sugars (Robards & Antolovich, 1997). Anthycyanin glycosylation usually occurs at position 3 with glucose, arabinose or galactose. Anthocyanins and their derivatives are responsible for the colouring of fruits of all kind whereas identified in leaf, seed and tiller (Wrolstad et al., 1990; Strack & Wray, 1993; Latti et al., 2009; Liu et al., 2013). According to Borowski et al. (2008) and Veberic et al. (2015), anthocyanins are located in the vacuoles of the granular form mainly in the outer tissue (the epidermis), whilesporadically present in the flesh. Nikkhah et al. (2010) suggest that there are 22 different anthocyanins known and that the six of them are most commonly present with food intake. The most common anthocyanins (Figure 2) are: cyanidin, peonidin, delphinidin, petunidin, malvidin and pelargonidin (Wrolstad, 1976; Nikkhah et al., 2010).

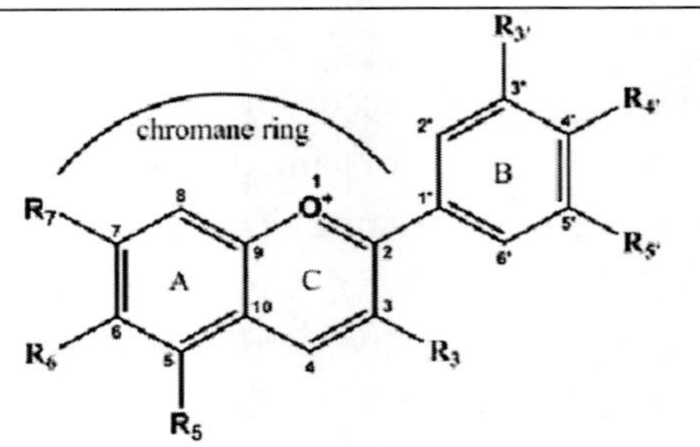

1	Pelargonidin (Pg)	R_3=OH; R_5=OH; R_6=H; R_7=OH; $R_{3'}$=H; $R_{4'}$=OH; $R_{5'}$=H
2	Cyanidin (Cy)	R_3=OH; R_5=OH; R_6=H; R_7=OH; $R_{3'}$=OH; $R_{4'}$=OH; $R_{5'}$=H
3	Delphinidin (Dp)	R_3=OH; R_5=OH; R_6=H; R_7=OH; $R_{3'}$=OH; $R_{4'}$=OH; $R_{5'}$=OH
4	Peonidin (Pn)	R_3=OH; R_5=OH; R_6=H; R_7=OH; $R_{3'}$=OMe; $R_{4'}$=OH; $R_{5'}$=H
5	Petunidin (Pt)	R_3=OH; R_5=OH; R_6=H; R_7=OH; $R_{3'}$=OMe; $R_{4'}$=OH; $R_{5'}$=OH
6	Malvidin (Mv)	R_3=OH; R_5=OH; R_6=H; R_7=OH; $R_{3'}$=OMe; $R_{4'}$=OH; $R_{5'}$=OMe

Figure 2. Chemical structure of the most common anthocyanins (Ananga et al., 2013).

Health Effects of Bioactive Compounds in Berries

In normal biological conditions, the oxygen molecule with non-enzymatic oxidation occasionally takes electrons from other molecules, which causes the formation of free radicals. Free radicals are permanent products of cellular metabolism (Benavente-García et al., 1997) with increased concentration in stress conditions (Nieman et al., 2002). Free radicals are formed as a result of air pollution, radiation, excessive exposure to the sun, stress, or because of excessive intake of industrial foods and foods rich in saturated fats. They lead to lipid peroxidation, damage of plasma membrane proteins and DNA (Berlett & Stadtman, 1997). Modern scientific studies indicate that oxidative stress is a major cause of various diseases, such as atherosclerosis, diabetes, malignant diseases, cardiovascular diseases and other chronic diseases. Therefore, the determination of antioxidant activity of berry fruits is significant in terms of assessment of the positive impact on human health (Sariburun et al., 2010). The largest share of biological effects of polyphenols can be attributed to their antioxidant activity. Narayana et al. (2001) and Liu (2003) define antioxidant activity as an ability to reduce free radicals and remove reactive oxygen species (ROS–Reactive Oxygen Species).

Due to the positive effect of fruit derived polyphenols on human health, there has been a steadily increasing interest in the consumption of fruit and fruit products. Berry varieties in comparison with other fruits and vegetables are characterized by the greatest antioxidant capacity (Lachman et al., 2000a, 2000b), which directly affects the removal of reactive oxygen species, inhibition of oxidation, and the development of the pathogenic bacteria (Kähkönen et al., 2001; Puupponen-Pimia et al., 2001). Phenolic compounds are characterized by anti-inflammatory, anti-viral, anti-microbial and anti-oxidative action, thus positively affecting human health (Reyes-Carmona et al., 2005). A large number of epidemiological studies indicate that eating fruits and vegetables helps to reduce the risk of some types of human cancers and cardiovascular diseases (Bazzano et al., 2002). Fruit consumptionalso has a positive impact on reduction in blood pressure, strengthening the immune system, detoxification of the body and reduction of inflammation (Sack & Kass, 1988; Ascherio et al. 1992).

Scalzo et al. (2005) suggest that antioxidant status is a more recent and significant parameter of the fruit quality. In the group of commercially important berry fruits, blackberry is characterized by the greatest antioxidant capacity of fruits due to the high content of phenols, or phenolic acids, anthocyanins, and other compounds of flavonoids (Pellegrini et al., 2003). Therefore, it should be more used in human diet (Sellappan et al., 2002; Zheng & Wang, 2003). Fruits of strawberry and blackberry have a high antioxidant activity *in vitro*, which is in a positive correlation to the polyphenolic compounds, such as anthocyanins. The antioxidant activity of anthocyanins is one of their most important biological properties relevant to human health (Heinonen et al., 1998; Wang & Jiao, 2000; Wang & Lin, 2000). Anthocyanins are potent antioxidants with the property of neutralizing free radicals, which can be attributed to the phenolic hydroxyl groups, located in their ring structure (Yoshiki et al., 1995; Wang et al., 1996; Rice-Evans et al., 1997; Wang et al., 1997).

Quantity of Bioactive Compounds in Berries

Berry fruits contain high concentrations of antioxidants, among which ascorbic acid (vitamin C) takes an important place. Nile & Park (2014), highlight that the content of vitamin C in berries depends upon a number of factors, such as the cultivar, growing technique, environmental conditions, fruit maturity, growth regions, storage time and conditions.

Pineli et al. (2011) found significantly lower content of vitamin C in fruits of strawberry cultivar 'Osogrande' compared to the results obtained by the Cordenunsi et al. (2002) and Pinto et al. (2007) in the same cultivar. According to Tulipani et al. (2011a, b) and Kafkas et al. (2006) the concentration of ascorbic acid is increased during fruit ripening. On the other hand, Ferreyra et al. (2007) have found a constant high content of vitamin C of the stage where the fruit is a white to increment the full maturity of the fruit (130 mg/100 g fw). Montero et al. (1996) have established that the highest level of vitamin C in fruits of strawberry was 35 days after fruit set, and it amounted to 100 mg/100 g fw. Cordenunsi et al. (2002), who state that the content, in full maturity stage, ranged from 40 to 85 mg per 100 g fw, have determined slightly lower values of vitamin C in fruits of strawberry. Olsson et al. (2004) reported a significant variation of the content of vitamin C in fruits of the strawberry cultivar 'Senga Sengana' depending on the site of cultivation. According to Moore et al. (2004), varying values in terms of the content of vitamin C in the fruit strawberry cultivar 'Bounty' have been influenced by years of research, a way of covering soil in the plantation and fertilizers applied. The same authors point out that the covering land with PE (polyethylene) foil had a positive effect on the concentration of vitamin C in fruits of strawberry, in contrast with straw mulching lands. On the other hand, Pantelidis et al. (2007) reported that the average content of vitamin C in berries ranges from 14.0 to 18.4 mg/100 g fw, while Benvenuti et al. (2004), and Romero-Rodriguez et al. (1992) shows a variation in the range 6.0 to 13.1 mg/100 g fw.

According to the research by Stanisavljević (1999), the content of vitamin C in the fruit of blackberries 'Čačanska bestrna' amounted to 17.2

mg/100 g fw, while Veberič et al. (2014) found a lower value (8.63 mg/100 g fw) for the same parameter. Variations in the content of vitamin C in strawberries and blackberries can be affected by the aforementioned factors, whereas the methods of extraction and analysis (applied to the fruit samples) may exert a significant effect.

Flavonoids and phenolic acids are two large and heterogeneous groups of biologically active non-nutritional phenolic compounds (Shahidi & Naczk, 1995). Phenolic acids (hydroxybensoic and hydroxycinnamic acid) are present in fruits in the form of esters, glycosides and amides. The most common phenolic acids from the group of the hydroxycinnamic acids, occurring in the fruits of berries are *p*-coumaric, caffeic and ferulic acid, and from hydroxybenzoic acids: *p*-hydroxybenzoic, gallic, ellagic, 3-4-dihydroxybenzoic and vanillic acid (Robards & Antolovich, 1997; Clifford, 1999; Tomas-Barberan & Clifford, 2000; Manach et al., 2004).

Ellagic acid seems to be very beneficial for human health (Losso et al., 2004; Bakkalbasi et al., 2009), but slightly present in a free form, usually after the acid hydrolysis reaction, as a decomposition product of ellagitannins (Beattie et al., 2005). Ellagitannins and derivatives thereof have been identified in certain small fruits only (raspberry, blackberry, arctic raspberry, strawberry) and some nuts (walnut, hazel) (Koponen et al., 2007). The concentration of ellagic acid in the berries is typically high, ranging from 16 to 207 mg/g dm (Williner et al., 2003). Mass et al. (1991) have found a significant amount of ellagic acid in the pulp of the green berries rather than in the pulp of red fruits. The highest content of the mentioned acid in blackberries has been found in the seed of the stone fruits (Siriwoharn & Wrolstad, 2004). By examining the structure of phenolic acids in the fruit of two blackberry cultivars ('Choctaw' and 'Kiowa'), Sellappan et al. (2002) identified gallic (6.42 and 4.12 mg/100 g fw in the order of sorts), caffeic (1.38 and 3 64 mg/100 g fw), *p*-coumaric (2.08 and 0.40 mg 100/g fw), ferulic (3.51 and 2.99 mg/100 g fw), and ellagic acid (33.81 and 30.01 mg/100 fw). The stated results confirm the findings of Siriwoharn & Wrolstad (2004) that the ellagic acid is a dominant phenolic acid in the fruits of blackberry.

By studying the changes in the content of phenolic compounds in fruits of six blackberry cultivars ('Black satin', 'Chester Thornless', 'Čačanska bestrna', 'Thornless Evergreen', 'Loch Ness' and 'Thornfree') during the freezing process, Veberič et al. (2014) found that the fruits of cultivar 'Čačanska bestrna' distinguished by a significantly lower content of ellagic acid (14.30 mg/kg fw), compared to other varieties studied (except in relation to the cultivar 'Chester Thornless'). Jakobek et al. (2007) have come up to similar conclusions. Namely, they state that the dominant phenolic acids present in the berries are ellagic (41 mg/kg fw) and *p*-coumaric (17 mg/kg fw). What is more, the results obtained by Häkkinen & Törrönen (2000) after studying the phenolic acid in 6 strawberry cultivars, indicate that ellagic acid accounted for the largest part of the total phenolic contents. Ellagic acid content ranged from 39.6 to 52.2 mg/100 g fw, while the content of *p*-coumaric acid ranged from 0.9 to 4,1 mg/100 g fw. The research results of these authors indicate that the largest variation between cultivars either in terms of the content of *p*-coumaric acid. On the other hand, Stöhr & Herrmann (1975) observed a significantly lower variation of the expressed under the influence of the genotype with respect to the content of *p*-coumaric acid (1−1.5 mg/100 g fw).

Red, blue and violet berry fruits represent the most important source of anthocyanins in relation to all edible plant species (Kähkönen et al., 2003). The structure and content of anthocyanins present in the fruits is caused by a number of factors, from the genetic variability to the factors of environment, such as light intensity, humidity, temperature, but also the application of fertilizers and pesticides, as well as infections with different pathogens (Kalt et al., 2001; Kähkönen et al., 2003; Hosseinian et al., 2007).

Crespo et al. (2009) stated that the profile of anthocyanins in strawberries is genetically determined trait rather than characteristic, influenced by environmental factors. The study of the stated authors suggest that cultivar 'Clery' has shown consistently high anthocyanin content reflecting the high stability of chemical composition of the fruit regardless of the area of cultivation. The content of total anthocyanins in the cultivar 'Clery' in this study ranged from 27.97 to 34.60 mg/100 g fw, which was 1.7 times more than the cultivar 'Anthony', which showed the lowest content of total anthocyanins in the fruit. Truax et al. (1994) suggest that the reason for a higher content of anthocyanins in strawberry fruit grown on polyethylene foil in relation to the strawberry, with straw as a mulching material, for about 10°C higher temperature under the foil than under the straw. Wang & Camp (2000) emphasize that the fruit of strawberries in peaks of day and night temperature (30/22°C) had the most intensely coloured skin and flesh. Schouten et al. (2003) suggest that the fruits of strawberries, which are more susceptible to damage, produce more anthocyanins. According to Rana (2001), the combined application of mineral nitrogen and biofertilizers have led to the increase of red coloured pigments in strawberry cultivars 'Chandler', in accordance with studies by Umar et al. (2009), which found an increase of anthocyanin content in the same cultivar using mineral nitrogen and *Azotobacter*.

Lopes da Silva et al. (2007) found 25 different anthocyanins in the fruits of five strawberry cultivars ('Eris', 'Oso Grande', 'Carisma', 'Tudnew' and 'Camaros'). Most anthocyanins were pelargonidin-aglycone, whereas the presence of some derivatives of cyaniding was found too. Glucose was the most common sugar, but rutinose, arabinose and rhamnose were found as well. In all five of the studied strawberry cultivars, pelargonidine-3-glucoside is the dominant anthocyanin followed by pelargonidine-3-rutinosid and cyanidine-3-glucoside. These three anthocyanins accounted for more than 95% of total anthocyanins in strawberries. In some earlier studies, cyanidine-3-glucoside was stated to be most abundant in fruits, as well as in other plant organs (Ishikura & Sugahara, 1979). The antioxidant efficacy in the prevention of oxidation of human low-density lipoprotein among the anthocyanins is the following: delphinidin > cyaniding > malvidin > pelargonidin (Satue-Garcia et al., 1997). Wang & Lin (2000) by examining the ORAC (Oxygen Radical Absorbance Capacity) at different stages of development of strawberry fruits, found a positive correlation between this parameter and the content of anthocyanins in the mature berries.

Different structure and a high content of anthocyanins present in the cells of blackberry fruits provide an intense, almost black colour of the fruit (Howard et al., 2012). Koca & Karadeniz (2009) by examining the structure of anthocyanin in the fruit of wild and seven cultivated varieties of blackberry, found the presence of 12 anthocyanins, of which 5 peaks were identified, and particularly for: cyanidin-3-glucoside (cyn-3-glu), cyanidin-3,5-diglucoside (cyn-3,5-di glu), peonidin-3-glucoside (peo-3-glu), pelargonidin-3-glucoside (plg-3-glu) and cyanidin-3-rutinoside (cyn-3-rut). The same authors emphasize that the content of cyaniding-3-glucoside in the fruit-berries ranged from 77.47 to 90.42%. Cho et al. (2004) obtained similar values of cyaniding-3-glucoside content in the fruits of 'Navajo' blackberry cultivar, while the value of the same parameter in the cultivar 'Arapaho' wasslightly lower. Fan-Chiang & Wrolstad (2005) identified in blackberry fruits cyanidin-3-glucoside, cyanidin-3-rutinoside, cyanidin-3-xyloside, cyanidin-3-glucoside, with apple acid and an acylated derivative of cyanidin-3-glucoside. According to Szajdek & Borowski (2008), the content of total anthocyanins in fruits of blackberry ranges from 134.6 to 152.2 mg/100 g fw. Similar values of total anthocyanins in the fruit of the two blackberry cultivars ('Choctaw', 'Kiowa') were obtained by Sellappan et al. (2002), while somewhat higher values were reported in three cultivars ('Chester Thornless', 'Hull Thornless' and 'Triple Crown') by Wang & Lin (2000).

According to the research by Stajčić et al. (2012), the content of anthocyanins in the fruits of black berry 'Čačanska bestrna' was 50.95 mg/100 g fw, while Veberič et al. (2014) found a slightly lower value (48.13 mg/100 g fw) for the same parameter in ecological conditions of the central Slovenia. By studying the antioxidant activity of fruits in conditions of various ultrasonic extraction treatments, Ivanović et al. (2014) have established that the total content of anthocyanin in the fruits of 'Čačanska bestrna' ranged from 1.15 to 1.30 mg/100 g dry matter.

Flavonols form one of the sub-group of flavonoids. Jakobek et al. (2007) outline that flavonol are present in lower concentrations in strawberry fruits compared to the phenolic acids. According to their presence in strawberry, kaempferol is in the first place (8 mg/kg fw), followed by quercetin (6 mg/kg fw). In addition, by measuring the content of flavonols Häkkinen et al. (1999) have found that the most abundant flavonol in strawberry fruits is kaempferol, and its concentration ranged from 0.2 to 0.9 mg/100 g fw, depending on the genotype. However, later research, carried out by Häkkinen et al. (1999) with the same strawberry cultivars indicated a significantly lower content of flavonols, and especially kaempferol, in fruits. The authors believe that this is due to a long period of the fruit storage in a frozen state prior to analysis. Pešaković & Milivojevic (2014) have reported a positive impact on biofertilizers on kaempferol content in the fruits of strawberry compared to mineral fertilizer. In the same study, there were no statistically significant differences in the contents of kaempferol among tested cultivars.

By examining the contents of flavonols in the fruits of four cultivars (‘Smoothstem’, ‘Satin White’, ‘Dirksen Thornless’ and ‘Hull Thornless’) and seven selections of thornless blackberry (C-33, C-55, C-60, C-57, C-62, C-58 and C-52), Bilyk & Sapers (1986) have found that the contents of kaempferol ranged from 0.6 to 2.6 mg/kg fw and the quercetin content of 5.2 to 35.4 mg/kg fw, with no myricetin detected in the samples of the studied cultivars and selections of blackberry. On the other hand, Milivojević et al. (2011) haven’t identified the quercetin in the fruits of wild and two commercial cultivars of blackberry (‘Thornfree’ and ‘Čačanska bestrna’), which is consistent with the allegations of Henning (1981), Bilyk & Sapers (1986) and Siriwoharn & Wrolstad (2004) on the absence of derivatives of quercetin in blackberry fruits. Based on the study of changes in the contents of anthocyanins and other compounds while freezing blackberries, Veberic et al. (2014) found that the content of flavonols in the fruits of ‘Čačanska bestrna’ was 12.62 mg/kg of fresh fruit mass. Milivojević et al. (2011) identified a lower content of kaempferol (1.55 mg/g fw), and myricetin (0.31 mg/g fw) in the fruits of ‘Čačanska bestrna’ compared to wild blackberries.

Influence of Strawberry Growing Technology on Bioactive Compound Improvement

Fertilization is one of the most important cultural practices used in modern strawberry production, including the use of organic and mineral fertilizers. An extensive body of research shows that the continuous use of mineral fertilizers leads to environmental contamination, with more than 50% of applied mineral fertilizers remaining unabsorbed, resulting in the loss of minerals, thus posing a serious threat to the environment. Therefore, the proper use and partial or complete substitution of mineral fertilizers with microbial inoculants i.e., biofertilizers can help overcome environmental problems caused by the overuse of mineral fertilizers.

Since strawberries are characterized by a high nutritional quality, reflecting in the contents of the fruit flavour, as well as in strong antioxidant capacity based on the presence of phenolic compounds and vitamin C, an important objective of this study is the examination of the influence of complex mineral fertilizers of various formulations and two biofertilizers (microbial biofertilizer based on a combination of bacteria of genera *Azotobacter*, *Azospirillum*, *Bacillus* and *Pseudomonas* and microbial biofertilizer based on bacteria of the genus *Klebsiella*) on variation in the content of said compounds. It will indicate the potential validity of this fertilizer application that have expressed the most positive impact.

In numerous studies, very different data for the content of vitamin C in the fruits of strawberries have been found. These variations can be the result of a variety of factors, including the cultivar, growing technique, environmental conditions, maturity, growth regions, and the length of the storage conditions (Nile & Park, 2014). Ferreyra et al. (2007) and Montero et al. (1996) noted considerably greater values of the vitamin C content in strawberries in relation to the values obtained in this chapter (Table 1). Based on the data presented in Table 1 we can see among analyzed strawberry cultivars that the significance of difference in respect of the vitamin C content has not been established. The results obtained in this chapter correspond to previously conducted researches by Cordenunsi et al. (2002), who stated that the content of vitamin C during strawberry ripening ranged from 40 to 85 mg per 100 g fw. Application of biofertilizer 2 had positive effect on the content of vitamin C in strawberries. These results are in line with the results by Moor et al. (2004), which concluded that the variation in the content of vitamin C in strawberries of the cultivar 'Bounty' was influenced by year of research, method of covering soil in the plantation and fertilizers applied.

Table 1. Effect of the cultivar and fertiliser type on the content vitamin C, total anthocyanin, total phenols and antioxidant capacity in strawberries

Factor		Vitamin C (mg/100 g fw)	Total anthocyanins (mg eq cyn-3-glu/100 g fw)	Total phenols (mg GA eq /100 g/fw)
Cultivar (A)	'Clery'	56.2 ± 3.6 a	31.6 ± 1.0 a	288.0 ± 22.2 a
	'Joly'	60.3 ± 3.4 a	24.5 ± 0.8 b	293.6 ± 16.9 a
	'Dely'	53.0 ± 1.6 a	32.9 ± 0.8 a	305.3 ± 18.7 a
Fertiliser (B)	MF	55.6 ± 3.1 b	30.9 ± 1.6 a	243.7 ± 10.2 c
	B1	58.3 ± 3.7 b	28.7 ± 1.3 a	302.1 ± 26.0 b
	B2	64.9 ± 2.8 a	29.6 ± 1.7 a	334.3 ± 19.5 a
	C	47.1 ± 1.9 c	29.4 ± 1.9 a	302.5 ± 19.9 b
ANOVA				
A		ns	*	ns
B		*	ns	*
A × B		*	*	*

MF – mineral fertilizer; B1 – biofertilzer 1; B2 – biofertilizer 2; C – control. The different lower-case letters in the columns indicate statistically significant differences among the mean values relative to cultivars and different fertiliser treatments at $P \leq 0.05$ level (LSD test). Asterisks in columns indicate significant differences for $P \leq 0,05$ (*) using F test. ns: not significant.

Interaction of cultivar and fertilizer type has shown a significant impact on the content of vitamin C in strawberries (Figure 3). In this regard, a significantly higher level of vitamin C in all fertlizer treatments including the control was found in the cultivar 'Joly' compared to cultivars

'Clery' and 'Dely'. Based on these data it can be concluded that the content of vitamin C is varietal-dependant characteristics (Tomić, 2016).

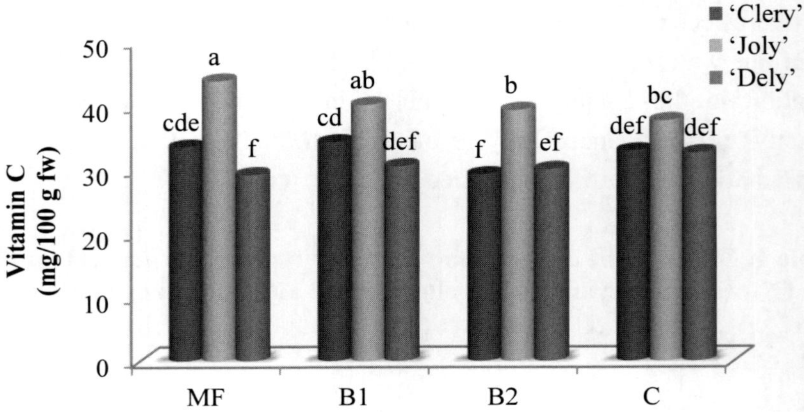

Figure 3. Interaction effect of the cultivar and fertiliser type on the content of vitamin C in strawberries (MF – mineral fertiliser; B1 – biofertilizer 1; B2 – biofrtilizer 2; C – control); the different letters indicate statistically significant differences at $P \leq 0.05$ level (LSD test).

Synthesis and distribution of phenolic compounds is conditioned by complex action from the outside (light, temperature and humidity), internal (genetic factors and hormonal status) (Strack, 1997), agro-ecological factors (Veberič et al., 2012), but also the growth condition and stage of fruit maturation (Murillo et al., 2012). In this regard, fertilizer treatments and the interaction combinations of cultivars and fertlizers conditioned significant variation regarding the total phenol content in strawberry (Table 1). Considerably lower total phenol content is registered in the treatment with the mineral fertilizer (243.7 mg gallic acid equivalent (GA eq)/100 g fw) in comparison to all other examined treatments. Significantly higher and also the highest total phenolic content is recorded in the treatment with biofertilizer 2 (334.3 mg GA eq/100 g fw). According to Kivijärvi (1999), synthesis of phenolic compounds might be accelerated in organic berry production where herbicides, pesticides, insecticides and fertilizers haven't been used. Similar to that Pešaković et al. (2016) explain the increase of total phenols and total anthocyanins in their study as a result of the biofertilizer applied.

Interaction effect of the cultivar and fertlizer on the total phenol content in strawberries is shown in Figure 4. Based on the data presented a significantly higher contents of total phenols in the fruit cultivars 'Joly' and 'Dely' can be seen in all examined fertilizer treatments compared to the control treatment, while in the cultivar 'Clery' no significant differences in total phenol content between the examined treatments were recorded.

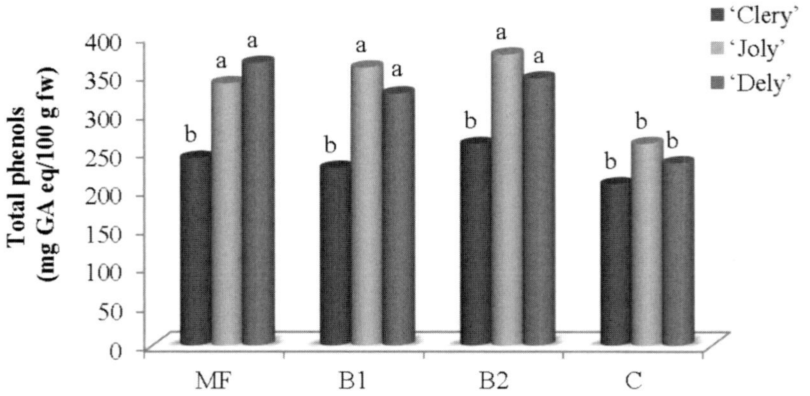

Figure 4. Interaction effect of the cultivar and fertiliser type on total phenols content in strawberries (MF – mineral fertiliser; B1 – biofertilizer 1; B2 – biofertilizer 2; C – control); the different letters indicate statistically significant differences at P≤ 0.05 level (LSD test).

Flavonoids are a special class of phenolic compounds of the basic skeleton C6-C3-C6, present in all plant organs (Milić et al., 2000). They are divided into twelve sub-class, comprising flavones, isoflavones, flavanones, flavonols, flavanols, flavenes, catechins, anthocyanidins, leucoanthocyanidin, chalcones, dihydrochalcone and aurons. Anthocyanins (anthos-flowers, kyanos-blue) are a subclass of these compounds with double significance, firstly, technological, due to the influence on the sensory characteristics of fruit products, and other biological activity, due to the effects on sensory characteristics of fruits and secondly, biological, due to the health effects among which the most important one is cardioprotective effect (De Pascual & Sanchez-Ballesta, 2008). These compounds contribute most to the total phenol content in berry fruits.

Kähkönen et al. (2003) and Hosseinian et al. (2007) highlight that the content and composition of anthocyanin in berry fruits, apart from a genotype (cultivar), is conditioned by light intensity, temperature, soil type, humidity, method of fertilizer and pesticide use and other stress factors. The proof of these are the results obtained in this study of Tomić (2016). The highest content of total anthocyanins is recorded in the cultivars ‘Clery’ and ‘Dely’ (31.6 and 32.9 mg equivalent cyanidin-3-glucoside (eq cyn-3-glu)/100 g fw, respectively), and somewhat lower value of the examined parameter in the cultivar ‘Joly’ (24.5 mg eq cyn-3-glu/100 g fw). The results obtained in this study correspond to previously conducted researches by Tomić et al. (2015), who stated that significantly higher values of vitamin C, total anthocyanins, total phenols, and antioxidative capacity were registered in ‘Joly’ compared to the other two examined cultivars.

The interaction effect of a cultivar and fertilizer on the content of total anthocyanin in strawberries is shown in Figure 5. A significantly higher content of total anthocyanins in the cultivar ‘Clery’ is registered in treatments with mineral fertilizer (36.7 mg eq cyn-3-glu/100 g fw) compared to all other treatments, while in the cultivar ‘Joly’ a significantly lower amount is recorded in the control treatment (22.8 mg eq cyn-3-glu/100 g fw), in response to the treatment with mineral fertilizer only (27.0 mg eq cyn-3-glu/100 g fw). Conversely, the cultivar ‘Dely’ had a significant increase in total anthocyanins in treatments with biofertilizer 2 and the control (31.7 and 35.6 mg eq cyn-3-glu/100 g fw, respectively) in comparison with other two treatments.

According to Veberič et al. (2015), the most common anthocyanidins in the fruits of garden strawberries are cyanidin and pelargonidin, though predominantly present in the form of glucoside, galactoside and glucoside. Lopes da Silva et al. (2007) have identified 25 different anthocyanins in the fruits of 5 strawberry cultivars and found that in all studied strawberry cultivars pelargonidin-3-glucoside is a predominant anthocyanin, followed by pelargonidin-3-rutinoside and cyanidin-3-glucoside. These three anthocyanins accounted for more than 95% of the total content of anthocyanins in strawberries. Our research also indicates the dominant presence of pelargonidin-3-glucoside compared to cyanidin-3-glucoside in the fruits of strawberries (Table 2).

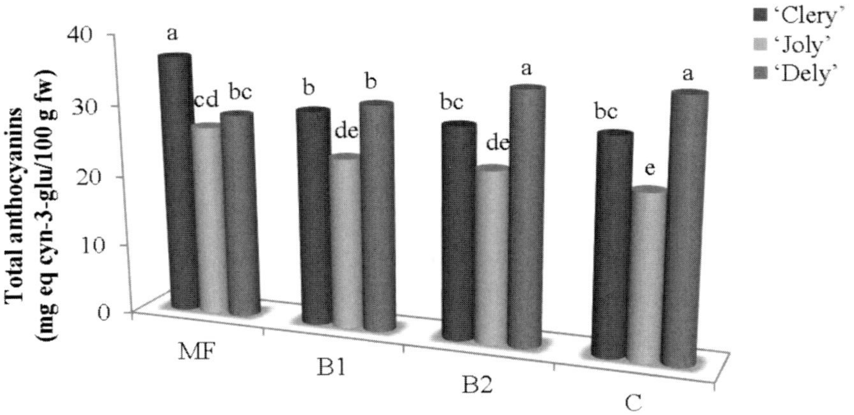

Figure 5. Interaction effect of the cultivar and fertilizer type on total anthocyanins content in strawberries (MF – mineral fertiliser; B1 – biofertilizer 1; B2 – biofrtilizer 2; C – control); the different letters indicate statistically significant differences at $P \leq 0.05$ level (LSD test).

The content of pelargonidin-3-glucoside was significantly higher in the fruit cultivars 'Clery' and 'Joly' (11.34 and 11.19 mg/100 g fw, respectively) in comparison with the cultivar 'Dely' (8.02 mg/100 g fw). After the treatment of plants using biofertilizer 1 and 2 a larger quantity of pelargonidin-3-glucoside was found in fruits (11.74 and 11.14 mg/100 g fw, respectively) compared to the plants treated with the mineral fertilizer and the control plants (8.76 and 9.10 mg/100 g fw, respectively). Interaction effects cultivar/fertilizer on the content of pelargonidin-3-glucoside in strawberries shown in Figure 6.

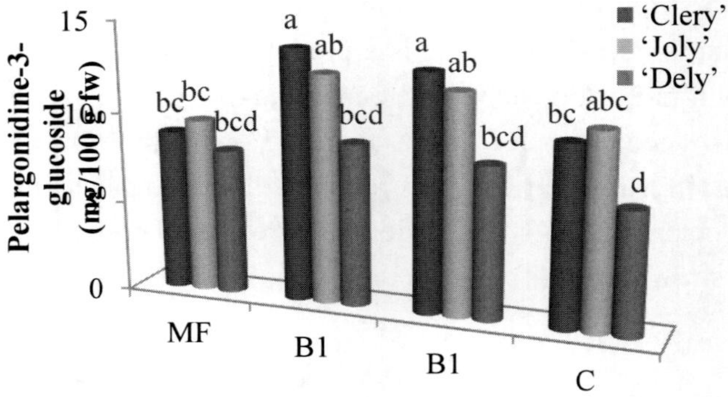

Figure 6. Interaction effect of the cultivar and fertiliser type on pelargonidine-3-glucoside content in strawberries (MF – mineral fertiliser; B1 – biofertilizer 1; B2 – biofrtilizer 2; C – control); the different letters indicate statistically significant differences at P ≤ 0.05 level (LSD test).

Table 2. Effect of the cultivar and fertilizer type on the content of flavanoids in strawberries

Factor		Flavanoids (mg/100 g fw)			
		Flavonols		Anthocyanins	
		Kaempferol	**Myricetin**	**Cyanidin-3-glucoside**	**Pelargonidin-3-glucoside**
Cultivar (A)	'Clery'	0.71 ± 0.02 a	0.93 ± 0.05 a	3.43 ± 0.21 b	11.34 ± 2.80 a
	'Joly'	0.68 ± 0.01 a	0.95 ± 0.11 a	4.29 ± 0.32 a	11.19 ± 1.86 a
	'Dely'	0.72 ± 0.04 a	0.94 ± 0.10 a	2.58 ± 0.46 c	8.02 ± 2.24 b
Fertilizer (B)	MF	0.79 ± 0.05 ab	0.88 ± 0.13 b	3.98 ± 0.25 a	8.76 ± 1.78 b
	B1	1.02 ± 0.02 a	0.97 ± 0.02 a	4.47 ± 0.27 a	11.74 ± 1.72 a
	B2	0.93 ± 0.02 a	0.91 ± 0.07 ab	3.22 ± 0.11 ab	11.14 ± 1.46 a
	C	0.66 ± 0.03 b	0.86 ± 0.06 b	2.97 ± 0.18 b	9.10 ± 1.07 b
ANOVA					
A		ns	ns	*	*
B		*	*	*	*
A × B		ns	ns	ns	*

MF – mineral fertilizer; B1 – biofertilzer 1; B2 – biofertilizer 2; C – control. The different lower-case letters in the columns indicate statistically significant differences among the mean values relative to cultivars and different fertiliser treatments at P ≤ 0.05 level (LSD test). Asterisks in columns indicate significant differences for P ≤ 0,05 (*) using F test. ns: not significant.

By analysing the presented data, it can be concluded that biofertilizer application caused a significant increase of the content of pelargonidin-3-glucoside in the cultivar 'Clery' (biofertilizer 1 – 13.77 mg/100 g fw and biofertilizer 2 – 12.98 mg/100 g fw) in comparison with treatments with mineral fertilizers and control treatment. The cultivars 'Joly' and 'Dely' did not differ significantly between treatment and the control treatment.

Significant differences in the content of cyanidin-3-glucoside are registered among all three studied cultivars. A significantly higher content of the compound was observed in the cultivar 'Joly' (4.29 mg/100 g fw) compared to other two studied cultivars. Treatments with mineral fertilizers and biofertilizer 1 caused a significant increase in the content of cyanidin-3-glucoside (3.98 and 4.47 mg/100 g fw, respectively) compared to the control (2.97 mg/100 g fw). The value obtained by using biofertilizer 2 was between stated values (3.22 mg/100 g fw). Interaction effect of cultivar and fertilizer on the content of cyanidin-3-glucoside in strawberries was not significant.

Flavonols are widespread in plants, primarily in the form of *O*-glycosides, among which most common are quercetin, myricetin, kaempferol and izorahmetin (Robards & Antolovich, 1997). The results obtained in this chapter confirm the claims of Jakobek et al. (2007), who state that flavonols are present in lower concentrations in strawberries compared to phenolic acids. In this study no influence of cultivar on the content of kaempferol was observed in strawberries. Strawberries treated with biofertlizers (1.02 and 0.93 mg/100 g fw, respectively) had a higher content of kaempferol compared to the control treatment fruits only. The results have substantial confirmations in the research by Pešaković & Milivojevic (2014) who have found a positive impact of biofertilizers on kaempferol content in strawberries compared to mineral fertilizer. In the same research, there were no statistically significant differences in the content of kaempferol among examined cultivars. The results obtained in study of Tomić (2016) are in compliance with previous Häkkinen & Törrönen (2000), who state that the contents of kaempferol in strawberries ranged from 0.2 to 0.9 mg/100 g fw depending on the genotype.

Larger contents of myricetin was in biofertilizer treatments (0.97 and 0.91 mg/100g of fw, respectively). Results obtained by Milivojević et al. (2011) indicate two to four times lower content of myricetin in fruits of two cultivars of garden and wild strawberry compared with the contents of kaempferol. One explanation for the low concentration of the mentioned flavonols is that a small number of compounds occur as free flavonoids-aglycones, while a number of them occur as flavonoids-glycosides, as a consequence of the basic compound to various monosaccharides or complex sugars (Milić et al., 2000).

Fenolic acids consist of phenolic nucleus and the side chain containing one (benzoic acid derivatives) or three (cinnamic acid derivatives) carbon atoms and include hydroxy and other functional derivatives of benzoic and cinnamic acids. The most abundant hydroxycinnamic acids, which occur in berries, are *p*-coumaric, caffeic and ferulic acid, hydroxybenzoic acid are: *p*-hydroxybenzoic acid, gallic, ellagic, 3-4-dihydroxybenzoic and vanillic acid (Robards & Antolovich, 1997; Clifford, 1999; Tomás-Barberan & Clifford, 2000; Manach et al., 2004). Milivojević (2008) highlights that ellagic acid represents a dimeric condensation product of gallic acid so its presence is usually in the form of ellagitannins. Free form of ellagic acid is rarely present in strawberries (da Silva Pinto et al., 2008). Pinto et al. (2007) state that the potential health effect of ellagitannins originate from strawberries, connected to the antiproliferative and *in vitro* inhibition of α-amylase, α-glucosidase inhibitors and ACE (angiotensin I-converting enzyme). The results of previous studies show that the content of free ellagic and gallic acid strawberries was dominant compared to ferulic and *p*-coumaric acid (Table 3). In the cultivar 'Dely' content of ellagic and gallic acid in the strawberries was significantly higher than the content of these acids in two other cultivars studied. Results of previous studies indicate the great variability value in the content ellagic acid in strawberries. Jakobek et al. (2007) state that the amount of ellagic acid in strawberries was 41 mg/kg fw, which is significantly lower in comparison to the values, set in this study. In contrast, by studying phenolic acid in the fruits of 6 strawberry cultivars Häkkinen & Törrönen (2000) have found that the largest part of the total phenolic contents was ellagic acid, the amount of which in fruits ranged from 39.6 to 52.2 mg/100 g fw, which is significantly higher than the values reported in this chapter.

Table 3. Effect of the cultivar and fertiliser type on content the phenolic acids in strawberries

Factor		Hydroxybenzoic acid (mg/100 g fw)		Hydroxycinnamic acids (mg/100 g fw)	
		Ellagic acid	**Gallic acid**	**Ferulic acid**	***p*-coumaric acid**
Cultivar (A)	'Clery'	9.15 ± 0.84 b	1.99 ± 0.56 b	0.23 ± 0.01 a	1.23 ± 0.15 a
	'Joly'	8.99 ± 1.10 b	2.01 ± 0.85 b	0.16 ± 0.03 b	1.09 ± 0.12 a
	'Dely'	12.22 ± 0.74 a	3.23 ± 0.64 a	0.20 ± 0.01 b	1.21 ± 0.14 a
Fertilizer (B)	MF	7.48 ± 0.67 c	2.29 ± 1.03 a	0.21± 0.02 a	1.19 ± 0.15 a
	B1	11.30 ± 0.57 a	2.45 ± 0.87 a	0.22 ± 0.02 a	1.22 ± 0.17 a
	B2	9.85 ± 0.46 b	2.34 ± 0.63 a	0.20 ± 0.03 ab	1.25 ± 0.19 a
	C	7.45 ± 0.48 c	2.23 ± 0.98 a	0.14 ± 0.01 b	1.18 ± 0.14 a
ANOVA					
A		*	*	*	ns
B		*	ns	*	ns
A × B		ns	ns	*	ns

MF – mineral fertilizer; B1 – biofertilzer 1; B2 – biofertilizer 2; C – control. The different lower-case letters in the columns indicate statistically significant differences among the mean values relative to cultivars and different fertiliser treatments at P≤ 0.05 level (LSD test). Asterisks in columns indicate significant differences for P ≤ 0.05 (*) using F test. ns: not significant.

Gallic acid has a significant antioxidant activity (Nile & Park, 2014), as well as its derivatives which are also powerful antioxidants with free hydroxyle with the property of neutralizing free radicals (Rice-Evans et al., 1997). Optimization of cultural practices, primarily fertilization may be one of the most effective ways to increase the phenolic content in strawberries (Anttonen et al., 2006). The same authors have shown that high levels of mineral fertilizers decreased content of ellagic acid and flavonols in strawberries. The results obtained correspond with the results of Tomić (2016), where the use of different biofertilizers caused increased content of phenolic acids, flavonols and anthocyanins in strawberries in relation to the use of mineral fertilizers. Namely, in this study, the application of biofertilizers 1 showed a stimulating effect on the content of ellagic acid in strawberries (11.30 mg/100 g fw). The obtained results were in accordance with the research by Kivijärvi (1999) who observed an increase in the content of phenolic component in the strawberries grown in organic farming system, without pesticides and fertilizer application. Similar to the above, in this chapter, a significantly higher content of gallic acid was registered in the control treatment without fertilization. No regularity in the content of gallic acid in the interaction cultivar/fertilizer was observed.

The influence of cultivar was significant in terms variability in the content of ferulic acid in the fruit. A significantly higher content of these acids was recorded in the cultivar 'Clery' (0.23 mg/100 g fw). On the other hand, positive effect of mineral fertilizers and biofertilizer 1 on the content of ferulic acid in strawberries has been noted. Interaction effect cultivar/fertilizer had a significant impact on the change of content of ferulic acid in strawberries, whereas the higher the content of the acid was recorded in the cultivar 'Clery' using a mineral fertilizer in relation to the other tested interaction effects tested (Figure 7). The high content of ferulic acid was recorded in the interaction cultivar 'Clery'/fertilizer but also in the interaction of the same cultivar with biofertilizers 1 and 2, among which no significant difference was observed.

No significant variation in the content of of *p*-coumaric acid under the influence of tested factors (cultivar and fertilizer), as well as under the influence of their interaction was observed. The content of *p*-coumaric acid ranged of from 1.09 to 1.25 mg/100 g fw. The amount of *p*-coumaric acid in the strrawberry quoted by Jakobek et al. (2007) was slightly higher (17 mg/kg fw) compared to the results recorded in this chapter. According to Häkkinen & Törrönen (2000), the average content of *p*-coumaric acid in strawberries ranges from 0.9 to 4.1 mg/100 g fw, which is in accordance with the values obtained in this study. Research results of the content of various phenolic acids in strawberries by the aforementioned authors suggest that the greatest variation among cultivars were regarding the content of *p*-coumaric acid. This was in contrast with the results obtained in this chapter, since the content did not change *p*-coumaric acid under the influence of the cultivar. On the other hand, Stöhr & Herrmann (1975) observed significantly less variation under the influence of genotype regarding the content of *p*-coumaric acid in strawberries (1–1.5 mg/100 g fw).

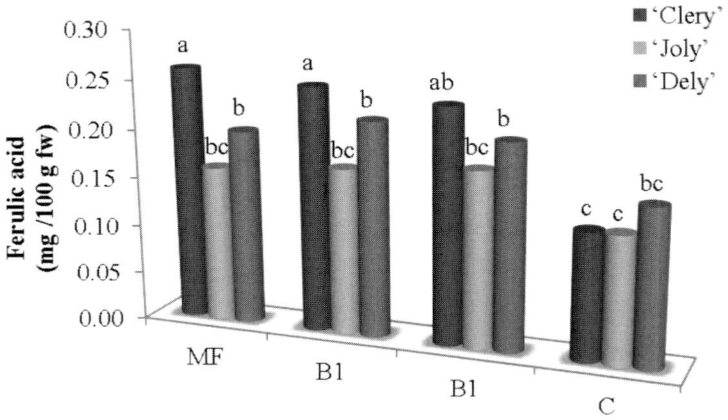

Figure 7. Interaction effect of the cultivar and fertiliser type on ferulic acid content in strawberries (MF – mineral fertiliser; B1 – biofertilizer 1; B2 – biofrtilizer 2; C – control); the different letters indicate statistically significant differences at $P \leq 0.05$ level (LSD test).

According to the quotes by Seeram et al. (2006) and Aaby et al.(2007), the fruits of strawberry contain, apart from vitamin C, large quantities of phenolic compounds, which exibited positive effect against free radicals *in vitro* examinations. Tulipani et al. (2008) emphasize that vitamin C account for 30–35% of antioxidant capacity of fruits in certain strawberry cultivars, whereas other specific compounds like pelargonidine-3-glucoside account for up to 25% antioxidant capacity of strawberries. Predominant phenolic compounds in strawberries in this chapter were pelargonidine-3-glucoside and ellagic acid, significantly contributing (along with vitamine C) to antioxidant capacity of the examined strawberry cultivar. Namely, significantly higher antioxidant capacity were found in 'Clery' and 'Joly' (1.14 and 1.03 mmol TE-Trolox equivalent/100 g fw, respectively) compared to 'Dely' (Table 4). Influence of fertlizer on strawberry antioxidant capacity was not recorded.

Table 4. Effect of the cultivar and fertiliser type on antioxidant capacity in strawberries

Factor		Antioxidant capacity (mmol TE·100 g/fw)
Cultivar (A)	'Clery'	1.14 ± 0.06 a
	'Joly'	1.03 ± 0.05 a
	'Dely'	0.75 ± 0.05 b
Fertiliser (B)	MF	0.97 ± 0.07 a
	B1	1.00 ± 0.09 a
	B2	0.90 ± 0.11 a
	C	1.02 ± 0.07 a
ANOVA		
A		*
B		ns
A × B		*

MF – mineral fertilizer; B1 – biofertilzer 1; B2 – biofertilizer 2; C – control. The different lower-case letters in the columns indicate statistically significant differences among the mean values relative to cultivars and different fertiliser treatments at P ≤ 0.05 level (LSD test). Asterisks in columns indicate significant differences for P ≤ 0,05 (*) using F test. ns: not significant.

By interaction effect analysis cultivar/fertilizer, significantly higher value of antioxidant capacity can be observed in 'Clery' applying biofertilier 1 compared to control (Figure 8).

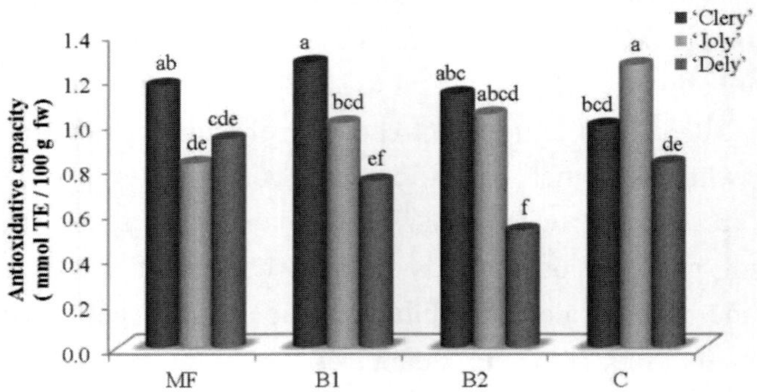

Figure 8. Interaction effect of the cultivar and fertiliser type on the antioxidative capacity in strawberries (MF – mineral fertiliser; B1 – biofertilizer 1; B2 – biofrtilizer 2; C – control); the different letters indicate statistically significant differences at P≤ 0.05 level (LSD test).

Biofertilizer application had a positive effect on the contents of most phenolic compounds analyzed in the fruit. Therefore, the use of microbiological fertilizers as supplements to mineral fertilizers or even their substitutes can be considered an appropriate practice to ensure safe strawberry fruit production, which has an indirect positive effect on the production characteristics of the tested strawberry cultivars.

INFLUENCE OF BLACKBERRY GROWING TECHNOLOGY ON BIOACTIVE COMPOUND IMPROVEMENT

The quality of blackberries as fruit species is fully accomplished only if the production is realized under conditions of properly selected localities, exposure time, altitude, an adequate type of soil, the cultivation techniques and proper, that is, timely execution of all agro-technical measures in accordance with biological requirements of the grown cultivar. Introduction of intensive blueberry cultivation systems, semi-enclosed, i.e., in rain-shield, prevents the harmful effects of abiotic factors and provides a continuous harvest, regardless of the outside conditions. Therefore, the main objective of this study was to determine the effect of rain-shield on the content of natural antioxidant compounds (vitamin C and phenolic compounds) in the blackberry cultivar 'Čačanska bestrna', having in mind their nutritional diet and medicinal value. Vitamin C is one of the antioxidants (Szajdek & Borowska, 2008) expressing its antioxidant activity mainly in synergism with flavonoids (Isler et al., 1988; Kähkönen et al., 2001). Karaklajić Stajić (2016) found significant differences in the content of vitamin C in blackberries using different cultivation techniques. Conversely, examined years showed no significant effect on the content of vitamin C. (Table 5).

Table 5. Effect of cultivation techniques and year on the content of vitamin C in blackberries 'Čačanska bestrna'

Treatment		Vitamin C (mg/100 g fw)
Cultivation techniques (A)		
Rain-shield		14.37 ± 0.24 a
Standard		12.55 ± 0.41 b
Year (B)		
2011.		13.16 ± 0.25 a
2012.		13.90 ± 0.69 a
2013.		13.33 ± 0.42 a
Cultivation techniques × Year (A × B)		
Rain-shield	2011.	13.42 ± 0.31 ab
	2012.	15.05 ± 0.42 a
	2013.	14.65 ± 0.14 a
Standard	2011.	12.90 ± 0.39 b
	2012.	12.75 ± 1.19 b
	2013.	12.01 ± 0.26 b
ANOVA		
A		*
B		ns
A × B		*

The different lower-case letters in the columns indicate statistically significant differences among the mean values relative to cultivation techniques and year at $P \leq 0.05$ level (LSD test). Asterisks in columns indicate significant differences for $P \leq 0.05$ (*) using F test. ns: not significant.

Namely, a higher content of vitamin C in blackberries was recorded in rain-shield system in accordance with the quotes by Pantelidis et al. (2007) and Hägg et al. (1995), who emphasized that the content of vitamin C in strawberries was conditioned by numerous factors including the applied growing technology. In addition, a variability in the content of vitamin C with standard cultivation techniques by experimental years was observed, which could be interpreted only in the context of significant influence of interaction effect of cultivation techniques and year (Figure 9). Pantelidis et al. (2007) found by studying antioxidant capacity, the content of phenols, anthocyanins and vitamin C in raspberries, blackberries, red currant and gooseberry determined the average content of vitamin C in blackberries 16.2 mg/100 g fw.

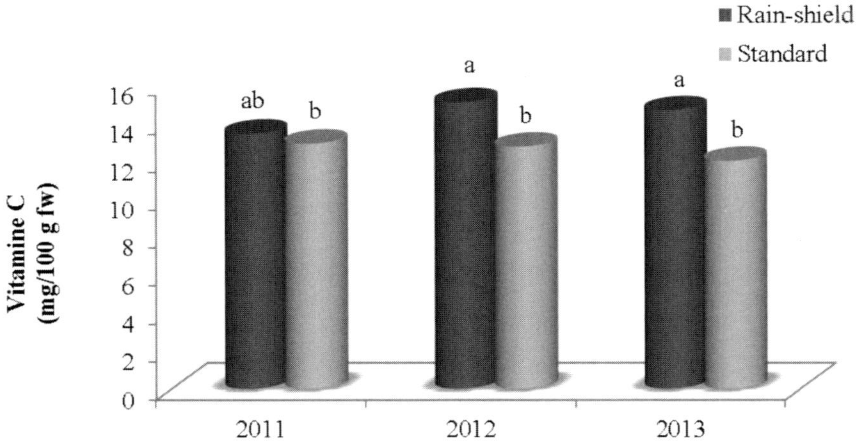

Figure 9. Interaction effect of the cultivation techniques and year on the content of vitamin C in blackberries; the different letters indicate statistically significant differences at $P \leq 0.05$ level (LSD test).

Similar research, related to the genera *Rubus*, *Ribes* and *Aronia* engaged in work by Benvenuti et al. (2004) indicated that the average content of the mentioned parameter in fruits of seven blackberry cultivars ('Black Diamond', 'Smoothstem', 'Thornless Boy Sembes', 'Darrow', 'Chester Thornless', 'Hull Thornless' and 'Black Satin') was 12.9 mg/100 g fw. Average values of vitamin C in this chapter per all treatments ranged from 12.01 to15.05% and were higher than the values stated by Veberic et al. (2014), for the cultivar 'Čačanska bestrna' in central Slovenia (8.63 mg/100 g fw), but also somewhat lower values for Cacak reached by Stanisavljević (1999) (17.2 mg/100 g fw).

The content of total phenols in the cultivar 'Čačanska bestrna', was considerably higher in rain-shield system related to the blackberry grown in open field, without rain-shield (Table 6). Observed by experimental years, higher values of total phenols were registered in 2011 and 2013 with the significance of differences in relation to 2012. Studying chemical properties of wild and commercial blackberries, Milivojević et al. (2010) found that the content of total phenols in 'Čačanska bestrna' were 1.74 mg eq GA/g fw which was lower than the results obtained in this work and comparable to the results reached by Stajčić et al. (2012) for the same parameter in the same cultivar (235.09 mg eq GA/100 g fw).

Table 6. Effect of cultivation techniques and year on the total phenols content in blackberries 'Čačanska bestrna'

Factor	Total phenols (mg eq GA/100 g fw)	Total anthocyanins (mg eq cyn-3-glu/100 g fw)
Cultivation techniques (A)		
Rain-shield	438.71 ± 15.02 a	70.98 ± 1.84 a
Standard	407.94 ± 11.49 b	70.40 ± 2.64 a
Year (B)		
2011.	451.24 ± 10.19 a	63.87 ± 1.80 b
2012.	362.65 ± 15.57 b	69.78 ± 2.36 b
2013.	456.09 ± 6.65 a	78.41 ± 2.37 a
ANOVA		
A	*	ns
B	*	*
A × B	*	*

The different lower-case letters in the columns indicate statistically significant differences among the mean values relative to cultivation techniques and year at $P \leq 0.05$ level (LSD test). Asterisks in columns indicate significant differences for $P \leq 0.05$ (*) using F test. ns: not significant.

Among phenolic compounds, predominant ones in blackberries are anthocyanidins, ellagic acid, flavonols and flavan-3-ols (Mosel & Herrmann, 1974; Schuster and Herrmann, 1985; Wald et al., 1986; Sellappan et al., 2002; Shahidi & Naczk, 2003). Values of total anthocyanins obtained in this chapter were approximately equal in both blackberry cultivation techniques confirming quotes by Milivojević (2008) that light is one of the most thoroughly studied factors of environment affecting metabolics of phenolic compounds, that is, stimulating the synthesis of flavonoids, especially anthocyanins and to a lesser degree flavonol glycosides. On the other hand, Karaklajić-Stajić et al. (2017) highlight that drastic chemical changes occur in the fruits during their ripening and the contribution of soluble solids content to stability of total anthocyanins was generally dependent on their levels in blackberries. Values of total anthocyanidins obtained in this study were higher in relation to the results reached by Stajčić et al. (2012), Ivanovic et al. (2014) and Veberic et al. (2014) for the same blackberry cultivar. The results of Karaklajić Stajić (2016) show that the ripeness stage significantly increase soluble solids content whereas total anthocyanins content has been noted in over ripe berries compared to fully ripe fruit resulting in lower rates of pigment degradation.

In the studied blackberry cultivar, quercetin, belonging to the group of flavonols and cyanidin-3-glucoside belonging to anthocyanins (Table 7) were registered. The obtained value of quercetin was higher in rain-shield, related to the standard cultivation techniques, although with no significance in differences. What is more, the content of the aforementioned compound did not vary significantly by experimental years neither, whereas the obtained values ranged from 0.25 to 0.33 mg/100 g fw. Values of quercetin content in fruits, obtained in this chapter were lower related to the values reached by Bilyk and Sapers (1986), by studying the contents of flavonols in the fruits of four cultivars and seven selections of thornless blackberry. On the other hand, Milivojević et al. (2011) point out that quercetin was not identified in blackberries 'Čačanska bestrna', which is, according to earlier studies by Henning (1981), Bilyk & Sapers (1986), Siriwoharn & Wrolstad (2004) the consequence of its presence in the form of glucoside.

Table 7. Effect of cultivation techniques and year on the flavonoids content in blackberries 'Čačanska bestrna'

Factor		Flavonoids (mg/100 g fw)	
		Flavonols	**Anthocyanins**
		Quercetin	**Cyanidin-3-glucoside**
Cultivation techniques (A)			
Rain-shield		0.31 ± 0.01 a	12.47 ± 1.25 a
Standard		0.29 ± 0.03 a	11.96 ± 1.12 a
Year (B)			
2011.		0.33 ± 0.02 a	6.28 ± 0.16 c
2012.		0.25 ± 0.01 a	12.51 ± 0.57 b
2013.		0.32 ± 0.04 a	17.86 ± 0.36 a
Factor		**Flavonoids (mg/100 g fw)**	
		Flavonols	**Anthocyanins**
		Quercetin	**Cyanidin-3-glucoside**
Cultivation techniques × Year (A × B)			
Rain-shield	2011.	0.35 ± 0.01 a	6.51 ± 0.31 a
	2012.	0.28 ± 0.01 a	12.14 ± 0.92 a
	2013.	0.30 ± 0.01 a	18.78 ± 0.16 a
Standard	2011.	0.30 ± 0.03 a	6.06 ± 0.04 a
	2012.	0.22 ± 0.02 a	12.88 ± 0.73 a
	2013.	0.35 ± 0.09 a	16.94 ± 0.44 a
ANOVA			
A		ns	ns
B		ns	*
A × B		ns	ns

The different lower-case letters in the columns indicate statistically significant differences among the mean values relative to cultivation techniques and year at $P \leq 0.05$ level (LSD test). Asterisks in columns indicate significant differences for $P \leq 0.05$ (*) using F test. ns: not significant.

Identification and classification of cyanidin-3-glucoside in the tested blackberry cultivar confirmed the quotes by Fan-Chiang & Wrolstad (2005), Jakobek et al. (2007) and Koca & Karadeniz (2009) that conjugated forms are predominant in the structure of blackberry anthocyanins, that is, glucoside and acylglucoside forms of cyanidine. Applied blackberry cultivation techniques did not cause significant difference in the content of cyanidin-3-glucoside in fruits, making the obtained values approximate, though somewhat higher in rain-shield system. During a three-year period, the values of cyanidin-3-glucoside content in fruits during the three-year period continually increased, so that the highest content of cyaniding-3-glucoside was in the third year, and a significant difference recorded between the years of testing. The incidence was probably a result of more favorable conditions for the synthesis of cyaniding-3-glucoside, considering that the structure and the content of anthocyanins in fruits caused by various factors, such as genetic variation, environmental factors, use of fertilizers and pesticides and pathogen infection (Kalt et al., 2001; Kähkönen et al., 2003; Cho, 2004; Hosseinian et al., 2007). According to the earlier investigation (Dugo et al., 2001; Fan-Chiang &Wrolstad, 2005; Stintzing et al., 2002, Veberic et al., 2014), in the structure of identified anthocyanins in blackberries, cyanidin-3-glucoside is predominant, which, according to Koca and Karadeniz (2009) acoounts for 83.95%, and according to Elisia et al. (2007) with 90.10%. Studying 14 different anthocyanins, Wang et al. (1997) found that cyanidin-3-glucoside had a highest antioxidant activity. Other anthocyanins, such as cyanidin-3-rutinoside, cyanidine-3-xyloside, malvidin-3-glucoside, an acylated derivative of cyanidine-3-glucoside were present in blackberries in lesser concentrations (Dugo et al., 2001).

In compliance with previous investigations (Dixon & Paiva, 1995; Deighton et al., 2000; Moyer et al., 2002; Määttä-Riihinen et al., 2004; Scalzo et al., 2005; Rutz et al., 2012), the results obtained in this chapter confirmed the influence of cultivation conditions, that is, abiotic factors and cultivation technique on changes in the content of phenolic acids and flavonoids in berry fruits. Blackberries in rain-shield system had a higher content of all identified hydroxybenzoic acids, except vanillic, and the significance of difference was registered in the content of ellagic and galic acid (Table 8). The highest content of 4-hydroxybenzoic, ellagic and gallic acids in fruits was found in 2013, with no significance in difference for the content of 4-hydroxybenzoic acid, by testing years. On the other hand, the highest content of protocatehic acid was found in 2012 whereas vanillic in 2011, with difference significance between testing years in the content of aforementioned acids. In all treatments, of 5 identified hydroxybenzoic acids, the highest content in fruits was registered in ellagic (6.77–16.17 mg/100 g fw), than gallic acid (3.44–8.00 mg/100 g fw), which is in accordance with the quotes by Siriwoharn & Wrolstad (2004) that ellagic acid is dominant fenolic acid in blackberries. Beattie et al. (2005) also emphasize that, in free form, ellagic acid is most commonly present with gallic acid, as a decomposition product of ellagitannins. Average values of the content of ellagic acid obtained in all treatments were higher than the values reached by Milivojević et al. (2011), studying chemical and antioxidant properties of cultivars and wild species of the genera *Fragaria* and *Rubus*, as well as the values quoted by Veberic et al. (2014) for the same cultivar.

P-coumaric, caffeic and ferulic acid from the goup of hydroxycinnamic acids were identified in this paper (Table 9). In rain-shield system grown blackberries, a higher content of the mentioned acids was obtained, with difference significance in the content *p*-coumaric acid only. The highest content of *p*-coumaric and caffeic acid was registered in 2013, ferulic acid in 2012, and difference in the content, between testing years were significant in all mentioned acids. An explanation of such incidence may be found in the fact that, apart from the genotype (Minoggio et al., 2002; Howard et al., 2003), the content of phenolic compound in fruits was also determined by temperature conditions during vegetation (Perez-Tello et al., 2001; Wang i Zheng, 2001), fruit maturation degree and abiotic factors (Reverberi et al., 2001; Kirakosyan et al., 2004).

Table 8. Effect of cultivation techniques and year on the hydroxybenzoic acids content in blackberries 'Čačanska bestrna'

Factor		Hydroxybenzoic acids (mg/100 g fw)				
		Protocatehic acid	4-hydroxybenzoic acid	Vanillic acid	Ellagic acid	Gallic acid
Cultivation techniques (A)						
Rain-shield		2.38 ± 0.09 a	0.60 ± 0.05 a	0.76 ± 0.08 a	11.36 ± 0.88 a	5.87 ± 0.42 a
Standard		2.14 ± 0.13 a	0.55 ± 0.04 a	0.89 ± 0.13 a	8.51 ± 0.46 b	4.25 ± 0.22 b
Year (B)						
2011.		1.91± 0.09 c	0.57 ± 0.07 a	1.35 ± 0.13 a	8.93 ± 0.37 b	4.11 ± 0.32 b
2012.		2.66 ± 0.16 a	0.56 ± 0.02 a	0.64 ± 0.04 b	7.31 ± 0.37 c	4.46 ± 0.14 b
2013.		2.21 ± 0.03 b	0.59 ± 0.06 a	0.50 ± 0.01 b	13.52 ± 0.82 a	6.60 ± 0.49 a
Cultivation techniques× Year (A × B)						
Rain-shield	2011.	2.09 ± 0.09 a	0.71 ± 0.11 a	1.08 ± 0.17 a	9.97 ± 0.39 c	4.79 ± 0.50 ab
	2012.	2.77 ± 0.15 a	0.55 ± 0.04 ab	0.70 ± 0.06 b	7.93 ± 0.49 d	4.81 ± 0.09 ab
	2013.	2.29 ± 0.02 a	0.54 ± 0.08 ab	0.52 ± 0.01 b	16.17 ± 0.29 a	8.00 ± 0.42 a
Standard	2011.	1.73 ± 0.13 a	0.44 ± 0.04 b	1.62 ± 0.11 a	7.89 ± 0.17 d	3.44 ± 0.16 b
	2012.	2.56 ± 0.28 a	0.57 ± 0.03 ab	0.59 ± 0.04 b	6.77 ± 0.49 e	4.12 ± 0.16 b
	2013.	2.14 ± 0.03 a	0.65 ± 0.09 ab	0.49 ± 0.22 b	10.87 ± 0.22 b	5.20 ± 0.33 ab
ANOVA						
A		ns	ns	ns	*	*
B		*	ns	*	*	*
A × B		ns	*	*	*	*

The different lower-case letters in the columns indicate statistically significant differences among the mean values relative to cultivation techniques and year at $P \leq 0.05$ level (LSD test). Asterisks in columns indicate significant differences for $P \leq 0.05$ (*) using F test. ns: not significant.

Table 9. Effect of cultivation techniques and year on the hydroxycinnamic acids content in blackberries 'Čačanska bestrna'

Factor		Hydroxycinamic acids (mg/100 g fw)		
		p-coumaric acid	Caffeic acid	Ferulic acid
Cultivation techniques (A)				
Rain-shield		2.28 ± 0.48 a	0.49 ± 0.02 a	0.42 ± 0.02 a
Standard		1.92 ± 0.45 b	0.47 ± 0.02 a	0.39 ± 0.01 a
Year (B)				
2011.		0.90 ± 0.04 b	0.41 ± 0.01 c	0.36 ± 0.02 b
2012.		0.64 ± 0.09 c	0.50 ± 0.02 b	0.47 ± 0.03 a
2013.		4.77 ± 0.12 a	0.55 ± 0.02 a	0.38 ± 0.01 b
Cultivation techniques × Year (A × B)				
Rain-shield	2011.	0.90 ± 0.06 b	0.42 ± 0.02 c	0.37 ± 0.04 a
	2012.	0.91 ± 0.07 b	0.48 ± 0.02 b	0.49 ± 0.05 a
	2013.	5.05 ± 0.08 a	0.58 ± 0.01 a	0.39 ± 0.01 a
Standard	2011.	0.89 ± 0.06 b	0.39 ± 0.01 c	0.34 ± 0.01 a
	2012.	0.37 ± 0.03 c	0.51 ± 0.02 ab	0.45 ± 0.03 a
	2013.	4.49 ± 0.15 a	0.52 ± 0.03 ab	0.37 ± 0.01 a
ANOVA				
A		*	ns	ns
B		*	*	*
A × B		*	*	ns

The different lower-case letters in the columns indicate statistically significant differences among the mean values relative to cultivation techniques and year at $P \leq 0.05$ level (LSD test). Asterisks in columns indicate significant differences for $P \leq 0.05$ (*) using F test. ns: not significant.

In respect of identified hydroxycinnamic acids in the fruits of the tested blackberry cultivars, we can observe the domination of *p*-coumaric acid, which is contrary to the quotes by Han et al. (2007), Pereira et al. (2009) and Dai et al. (2010) that in plant cells from the group of hydroxycinnamic acids, the most common is caffeic acid. Analyzing the group of berries Pantelidis et al. (2007) have found that the fruits of blackberry are characterized by the greatest antioxidant capacity resulting from the high content of phenolic compounds.

The values of blackberry antioxidant capacity obtained in study of Karaklajić Stajić (2016) were higher in rain-shield system, which was expected, considering the values of the total phenols and anthocyanins were higher as well in the mentioned system (Table 10). The relationship between antioxidant capacity and the content of total phenols and anthocyanins in the fruits of berries has been confirmed by numerous studies (Wang & 38; Lin, 2000; Sellappan et al., 2002; Reyes-Carmona et al., 2005; Milivojević et al., 2011) indicating the fact that the parameter 'total phenols' can serve as an indicator of antioxidant capacity (Milivojević et al., 2010). Regression-correlation analysis in our study didn't show a correlation between the content of total phenols and anthocyanins and the antioxidant capacity of the blackberry fruit cultivated under rain-shield system, contrary to the standard system where a positive linear relationship between the parameters mentioned above in a three-year period was observed (Table 11). Namely, a positive linear correlation was medium strong with approximately equal values of Pierson correlation coefficient between total phenols and anthocyanins and antioxidant capacity. Boyer & Liu (2004) have established that the fruits of apple exposed to sunlight were characterized by the higher content of anthocyanins and quercetin glycosides in relation to the fruits grown in overshadowing conditions. Contrary to the results presented in this chapter, higher values of correlation coefficient were found by Wang & Lin (2000) in cultivars 'Chester Thornless', 'Hulll Thornless' and 'Triple Crown', Sellapan et al. (2002) in cultivars 'Choctaw' and 'Kiowa' and Reyes-Carmona et al. (2005) in cultivars 'Brazos', 'Tupi', 'Comanche Wild', 'Evergreen', 'Marion' and 'Siskiyou'. In addition to that, the values of correlation coefficient between the content of total phenols and the fruit antioxidant capacity presented in Table 11 were lower, compared to the results found by Milivojević et al. (2010) and Milivojević et al. (2011) for the same cultivar.

Table 10. Effect of cultivation techniques and year on the antioxidant capacity in blackberries 'Čačanska bestrna'

Factor	Antioxidant capacity (mmol TE/100 g fw)
Cultivation techniques (A)	
Rain-shield	3.28 ± 0.08 a
Standard	2.72 ± 0.11 b
Year (B)	
2011.	2.90 ± 0.07 b
2012.	2.69 ± 0.15 c
2013.	3.40 ± 0.11 a
ANOVA	
A	*
B	*
A × B	*

The different lower-case letters in the columns indicate statistically significant differences among the mean values relative to cultivation techniques and year at $P \leq 0.05$ level (LSD test). Asterisks in columns indicate significant differences for $P \leq 0.05$ (*) using F test.

Table 11. Values of Pierson linear correlation coefficient between the content of total anthocyianin, total phenols and antioxidant capacity of the blackberries 'Čačanska bestrna'

Rain-shield system					
Parameter	**AC***	**TA**	**Parameter**	**AC**	**TP**
AC	/		AC	/	
TA	0.43	/	TP	0.34	/
Standard system					
Parameter	**AC**	**TA**	**Parametar**	**AK**	**TP**
AC	/		AC	/	
TA	0.55*	/	TP	0.51*	/

AC* – antioxidant capacity;

TA – total anthocyanins;

TP – total phenols;

Indicated values of correlation coefficient are statistically significant for $P \leq 0,01$.

The rain-shield system resulted in higher values for all parameters of the phenolic compounds and vitamin C, except the content of vanillic acid.

The results confirmed the presence of the flavonol quercetin and dominant presence of anthocyanin cyaniding-3-glucoside. Among phenolic compaunds detected in fruit of the tested cultivar, gallic and ellagic acid were found in high concentrations. The rain-shield cultivation system induced a significantly higher content of total phenols and, accordingly, significantly higher total antioxidant capacity (3.28 mmol TE/100 g fw) of the fruit.

ACKNOWLEDGMENTS

This study is the part of the project No. 31093: 'Content of bioactive components in small and stone fruits as affected by cultivar specificities and growing conditions, and obtaining biologically valuable products by improved and newly developed technologies', financed by Ministry of education, science and technological development of the Republic of Serbia. We hereby express our sincere gratitude for the support.

REFERENCES

Aaby, K., Ekeberg, D. & Skrede, G. (2007). Characterization of phenolic compounds in strawberry (Fragaria × ananassa) fruits by different HPLC detectors and contribution of individual compounds to total antioxidant capacity. *Journal of Agricultural and Food Chemistry*, *55*, 4395–4406.

Ananga, A., Georgiev, V. & Tsolova, V. (2013). Manipulation and engineering of metabolic and biosynthetic pathway of plant polyphenols. *Current Pharmaceutical Design*, *19*, 6186–6206.

Anttonen, M. J., Hoppula, K. I., Nestby, R., Verheul, M. J. & Karjalainen, R. O. (2006). Influence of fertilization, mulch color, early forcing, fruit order, planting date, shading, growing environment, and genotype on the contents of selected phenolics in strawberry (Fragaria × ananassa

Duch.) fruits. *Journal of Agricultural and Food Chemistry, 54*, 2614−2620.

Ascherio, A., Rimm, E. B., Giovannucci, E. L., Colditz, G. A., Rosner, B., Willett, W. C. F. & Stampfer, M. J. (1992). A prospective study of nutritional factors and hypertension among US men. *Circulation, 86*, 1475−1484.

Bakkalbasi, E., Mentes, Ö. & Artik, N. (2009). Food ellagitannins – occurrence, effects of processing and storage. *Food Science and Nutrition, 49*, 283–298.

Bazzano, L. A., He, J., Ogden, L. G., Loria, C. M., Vupputuri, S., Myers, L. & Whelton, P. K. (2002). Fruit and vegetable intake and risk of cardiovascular disease in US adults: the first national health and nutrition examination survey epidemiologic follow-up study. *The American Journal of Clinical Nutrition*, *76*, 93−99.

Beattie, J., Crozier, A. & Duthie, G. (2005). Potential health benefits of berries. *Current Nutrition and Food Science*, *1*, 71–86.

Benavente-García, O., Castillo, J., Marin, F. R., Ortuño, A. & Del Río, J. A. (1997). Uses and properties of citrus flavonoids. *Journal of Agricultural and Food Chemistry*, *45*, 4505−4515.

Benvenuti, S., Pellati, F., Melegar, M. & Bertelli, D. (2004). Polyphenols, anthocyanins, ascorbic acid, and radical scavenging activity of Rubus, Ribes, and Aronia. *Journal of Food Science*, *69*, 3, 164–169.

Berlett, B. S. & Stadtman, E. R. (1997). Protein oxidation in aging, disease and oxidative stress. *The Journal of Biological Chemistry*, *272*, 20313–20316.

Bilyk, A. & Sapers, G. M. (1986). Varietal differences in the quercetin, kaempferol, and myricetin contents of highbush blueberry, cranberry, and thornless blackberry fruits. *Journal of Agricultural and Food Chemistry*, *34*, 585–588.

Borowski, J., Szajdek, A., Borowska, E. J., Ciska, E. & Zieliński, H. (2008). Content of selected bioactive components and antioxidant properties of broccoli (Brassica oleracea L.). *European Food Research and Technology*, *226*, 3, 459–465.

Boyer, J. & Liu, R. H. (2004). Apple phytochemicals and their health benefits. *Nutrition Journal*, *3*, 1−15.

Bravo, L. (1998). Polyphenols: Chemistry, dietary sources, metabolism and nutritional significanse. *Nutrition Reviews*, *56*, 11, 317–333.

Caridi, D., Trenerry, V. C., Rochfort, S., Duong, S., Laugher, D. & Jones, R. (2007). Profiling and quantifying quercetin glucosides in onion (Allium cepa L.) varieties using capillary zone electrophoresis nad high performance liquid chromatography. *Food Chemistry*, *105*, 691–699.

Castro, E. (2009). Vocabulário de Foucault: um percurso pelos seus temas, conceitos e autores. [Vocabulary of Foucault: a journey through its themes, concepts and authors.] *Autêntica Editora.*

Cheynier, V. (2012). Phenolic compounds: from plants to foods. *Phytochemistry Reviews*, *11*, 2-3, 153–177.

Cho, M. J., Howard, L. R., Prior, R. L. & Clark, J. R. (2004). Flavonoid glycosides and antioxidant capacity of various blackberry, blueberry and red grape genotypes determined by high-performance liquid chromatography/mass spectrometry. *Journal of the Science of Food and Agriculture*, *84*, 1771–1782.

Clifford, M. N. (1999). Chlorogenic acids and other cinnamates–nature, occurrence and dietary burden. *Journal of the Science of Food and Agriculture*, *79*, 362−372.

Cordenunsi, B. R., Nascimento, J. R. O., Genovese, M. I. & Lajolo, F. M. (2002). Influence of cultivar on quality parameters and chemical composition of strawberry fruits grown in Brazil. *Journal of Agricultural and Food Chemistry*, *50*, 2581–2586.

Cos, P., Ying, L., Calomme, M., Hu, J. P., Cimanga, K., Poel, B. V., Pieters, L., Vlietinck, A. J. & Berghe, D. V. (1988). Structure-activity relationship and classsification of flavonoids as inhibitors of xanthine oxidase and superoxide scavengers. *Journal of Natural Products*, *61*, 71–76.

Crespo, P., Bordonaba, J. G., Terry, L. A. & Carlen, C. (2009). Characterisation of major taste and health-related compounds of four strawberry genotypes grown at different Swiss production sites. *Food Chemistry*, *122*, 16–24.

Dai, J. & Mumper, R. J. (2010). Plant phenolics: extraction, analysis and their antioxidant and anticancer properties. *Molecules*, *15*, 10, 7313–7352.

Da Silva Pinto, M., Kwon, Y. I., Apostolidis, E., Lajolo, F. M., Genovese, M. I. & Shetty, K. (2008). Functionality of bioactive compounds in Brazilian strawberry (Fragaria x ananassa Duch.) cultivars: evaluation of hyperglycemia and hypertension potential using *in vitro* models. *Journal of Agricultural and Food Chemistry*, *56*, 4386–4392.

Deighton, N., Brennan, R., Finn, C. & Davies, H. V. (2000). Antioxidant properties of domesticated and wild Rubus species. *Journal of the Science of Food and Agriculture*, *80*, 1307–1313.

De Pascual-Teresa, S. & Sanchez-Ballesta, M. T. (2008). Anthocyanins: from plant to health. *Phytochemistry Reviews*, *7*, 2, 281–299.

Ding, M., Feng, R., Wang, S. Y., Bowma, L., Lu, Y., Qian, Y., Castranova, V., Jiang, B. H. & Shi, X. (2006). Cyanidin-3-glucoside, a natural product derived from blackberry, exhibits chemopreventive and chemotherapeutic activity. *Journal of Biological Chemistry*, *281*, 25, 17359–17368.

Dixon, R. A. & Paiva, N. L. (1995). Stress-induced phenylpropanoid metabolism. *Plant Cell*, *7*, 1085–1097.

Dugo, P., Mondello, L., Errante, G., Zappia, G. & Dugo, G. (2001). Identification of anthocyanins in berries by narrow-bore high-performance liquid chromatography with electrospray ionization detection. *Journal of Agricultural and Food Chemistry*, *49*(8), 3987-3992.

Elisia, I., Hu, C., Popovich, D. G. & Kitts, D. D. (2007). Antioxidant assessment of an anthocyanin-enriched blackberry extract. *Food Chemistry*, *101*, 1052–1058.

Fan-Chiang, H. J. & Wrolstad, R. E. (2005). Anthocyanin pigment composition of blackberries. *Journal of Food Science*, *70*, 198–202.

Ferreyra, R. M., Viña, S. Z., Mugridge, A. & Chaves, A. R. (2007). Growth and ripening season effects on antioxidant capacity of strawberry cultivar Selva. *Scientia Horticulturae*, 112, 27–32.

Fiorani, M., De Sancitis, R., De Bellis, R. & Dacha, M. (2002). Intracellular flavonoids as electron donors forextracellular ferricyanide reduc tion in human erythrocytes. *Free Radical Biology & Medicine*, *32*, 64–72.

Hägg, M., Ylikoski, S. & Kumpulainen, J. (1995). Vitamin C content in fruits and berries consumed in Finland. *Journal of Food Composition and Analysis*, *8*, 12–20.

Häkkinen, S. H., Kärenlampi, S. O., Heinonen, I. M., Mykkänen, H. M. & Törrönen, A. R. (1999). Content of the flavonols quercetin, myricetin, and kaempferol in 25 edible berries. *Journal of Agricultural and Food Chemistry*, *47*, 2274−2279.

Häkkinen, S. H. & Törrönen, A. R. (2000). Content of flavonols and selected phenolic acids in strawberries and Vaccinium species: influence of cultivar, cultivation site and technique. *Food research international*, *33*, 517−524.

Haminiuk, C. W., Maciel, G. M., Plata-Oviedo, M. S. & Peralta, R. M. (2012). Phenolic compounds in fruits–an overview. *International Journal of Food Science & Technology*, *47*, 10, 2023−2044.

Han, X., Shen, T. & Luo, H. (2007). Dietary Polyphenols and their biological significance. *International Journal of Molecular Sciences*, *8*, 950–988.

Harborne, J. B. & Baxter, H. (1999). *The Handbook of Natural Flavonoids*. John Wiley and Sons, Chichester, UK.

Heinonen, I. M., Meyer, A. S. & Frankel, E. N. (1998). Antioxidant activity of berry phenolics on human low-density lipoprotein and liposome oxidation. *Journal of Agricultural and Food Chemistry*, *46*, 4107–4112.

Henning, W. (1981). Flavonol glycosides of strawberries (Fragaria × ananassa Duch.), raspberries (Rubus idaeus L.) and blackberries (Rubus fruticosus L.). *Zeitschrift fur Lebensmittel Untersuchung und Forschung*, *173*, 3, 180–187.

Herrmann, K. (1976). Flavonols and flavones in food plants: a review. International *Journal of Food Science & Technology*, *11*, 433–448.

Hosseinian, F. S., Li, W. & Beta, T. (2007). Measurement of anthocyanins and other phytochemicals in purple wheat. *Food Chemistry*, *109*, 916–924.

Howard, L. R., Clark, J. R. & Brownmiller, C. (2003). Antioxidant capacity and phenolic content in blueberries as affected by genotype and growing season. *Journal of the Science of Food and Agriculture*, *83*, 12, 1238–1247.

Howard, L. R., Prior, R. L., Liyanage, R. & Lay, J. O. (2012). Processing and storage effect on berry polyphenols: Challenges and implications for bioactive properties. *Journal of Agricultural and Food Chemistry*, *60*, 6678–6693.

Isler, O., Brubacher, G. & Kiss, J. (1988). Skorbut, vitamin C and bioflavonoide. In: '*Vitamine II: Wasserlosliche vitamine*'. Isler, O., Brubacher, G., Ghisla, S., Krautler, B. (eds.), Thiem, Stuttgart, pp. 390–395.

Ishikura, N. & Sugahara, K. (1979). A survey of anthocyanins in fruits of some angiosperms, II. *The Botanical Magazine*, *92*, 157–161.

Ivanović, J., Tadic, V., Dimitrijevic, S., Stamenic, M., Petrovic S. & Zizovic, I. (2014). Antioxidant properties of the anthocyanin-containing ultrasonic extract from blackberry cultivar 'Čačanska Bestrna'. *Industrial Crops and Products*, *53*, 274–281.

Jakobek, L., Šeruga, M., Medvidović-Kosanović, M. & Jovanović, I. N. (2007). Anthocyanin content and antioxidant activity of various red fruit juices. *Deutsche Lebensmittel Rundshau*, *103*, 58–64.

Kafkas, E., Koşar, M., Türemiş, N. & Başer, K. H. C. (2006). Analysis of sugars, organic acids and vitamin C contents of blackberry genotypes from Turkey. *Food Chemistry*, *97*, 732–736.

Kähkönen, M. P., Hopia, A. I. & Heinonen, M. (2001). Berry phenolics and their antioxidant activity. *Journal of Agricultural and Food Chemistry*, *49*, 8, 4076–4082.

Kähkönen M. P., Heinamaki J., Ollilainen V. & Heinonen M. (2003). Berry anthocyanins: isolation, identification and antioxidant activities. *Journal of the Science of Food and Agriculture*, *83*, 14, 1403–1411.

Kalt, W., Ryan, D. A., Duy, J. C., Prior, R. L., Ehlenfeldt, M. K. & Vander Kloet, S. P. (2001). Interspecific variation in anthocyanins, phenolics, and antioxidant capacity among genotypes of highbush and lowbush blueberries (Vaccinium section cyanococcus spp.). *Journal of Agricultural and Food Chemistry*, *49*, 10, 4761–4767.

Karaklajić-Stajić, Ž. (2016). Effect of rain-shield cultivation system on biological and production characteristics and fruit quality of blackberry cv. 'Čačanska bestrna' (*Rubus* subg. *Rubus* Watson). University of Belgrade, Faculty of Agriculture, 1–148.

Karaklajić-Stajić, Ž., Nikolić, M., Tomić, J., Miletić, R. & Pešaković, M. (2017). Influence of ripeness stage on the rate of pigment degradation in blackberry fruits during frozen storage. *European Journal of Horticultural Science*, *82*, 4, 198–203.

Katalinić, V., Smole Možina, S., Skroza, D., Generalić, I., Abramović, H., Miloš, M., Ljubenkov, I., Piskernik, S., Pezo, I., Terpinc, P. & Boban, M. (2010). Polyphenolic profile, antioxidant properties and antimicrobal activity of grape skin extracts of 14 Vitis vinifera varieties grown in Dalmatia (Croatia). *Food Chemistry*, *119*, 715–723.

Kirakosyan, A., Kauffman, P., Warber, S., Zick, S., Aaronson, K., Bolling, S. & Chanc, S. C. (2004). Applied environmental stresses to enhance the levels of polyphenolics in leaves of hawthorn plants. *Physiologia Plantarum*, *121*, 182–186.

Kivijärvi, P. (1999). *Organic currant production in the South Savo region in Finland.* Nordisk Jordbrugsforskning, Denmark.

Klein, B. P. & Kurilich, A. C. (2000). Processing effects on dietary antioxidants from plant foods. *HortScience*, *35*, 580−584.

Koca, I. & Karadeniz, B. (2009). Antioxidant properties of blackberry and blueberry fruits grown in the Black sea region of Turkey. *Scientia Horticulturae*, *121*, 447–450.

Koponen, J. M., Happonen, A. M., Mattila, P. H. & Törrönen, A. R. (2007). Contents of anthocyanins and ellagitannins in selected foods consumed in Finland. *Journal of Agricultural and Food Chemistry*, *55*, 4, 1612–1619.

Lachman, J., Orsak, M. & Pivec, V. (2000a). Antioxidant contents and composition in some vegetables and their role in human nutrition. *Horticultural Science*, *27*, 2, 65–78.

Lachman, J., Orsak, M. & Pivec, V. (2000b). Antioxidant contents and composition in some fruits and their role in human nutrition. *Horticultural Science*, *27*, 3, 103–117.

Larson, R. A. (1988). The antioxidants of higher plants. *Phytochemistry*, *4*, 969−978.

Lattanzio, V., Di Venere, D., Linsalata, V., Bertolini, P., Ippolito, A. & Salerno, M. (2001). Low temperature metabolism of apple phenolics and quiescence of Phlyctaena vagabunda. *Journal of Agricultural of Food Chemistry*, *49*, 5817–5821.

Latti, A. K., Kainulainen, P. S., Hayirlioglu-Ayaz, F. A. & Riihinen, K. R. (2009). Characterization of anthocyanins in Caucasian blueberries (Vaccinium arctostaphylos L.) native to Turkey. *Journal of Agricultural and Food Chemistry*, *57*, 12, 5244–5249.

Liu, R. H. (2003). Health benefits of fruit and vegetables are from additive and synergistic combinations of phytochemicals. *American Journal of Clinical Nutrition*, *78*, 517–520.

Liu, Y., Che, F., Wang, L., Meng, R., Zhang, X. & Zhao, Z. (2013). Fruit coloration and biosynthesis after bag removal in non-red and red apples (Malus × domestica Borkh.). *Molecules*, *18*, 1549–1563.

Lopes da Silva, M. F., Escribano-Bailón, M. T. & Santos-Buelga, C. (2007). Stability of pelargonidin 3-glucoside in model solutions in the presence and absence of flavanols. *American Journal of Food Technology*, *7*, 602−617.

Losso, J. N., Bansode, R. R., Trappey, A., Bawadi, H. A. & Truax, R. (2004). *In vitro* antiproliferative activities of ellagic acid. *Journal of Nutritional Biochemistry*, *15*, 11, 672–678.

Määttä-Riihinen, K. R., Kamal-Eldin, A. & Törrönen, A. R. (2004). Identification and quantification of phenolic compounds in berries of

Fragaria and Rubus species (family Rosaceae). *Journal of Agriculture and Food Chemistry*, *52*, 6178–6187.

Macheix, J. J., Fleureit, A. & Billot, J. (1990). Fruit phenolics. CRC Press, Boca Raton, Florida, USA.

Maestri, D. M., Nepote, V., Lamarque, A. I. & Zygadlo, J. A. (2006). Natural products as antioxidants. *Phytochemistry: Advances in Research*, *37*, 661, 105–135.

Manach, C., Scalbert, A., Morand, C. & Jimenez, L. (2004). Polyphenols: food sources and bioavailability. *American Journal of Clinical Nutrition*, *79*, 727–747.

Mass, J. L., Galletta, G. J. & Stoner, G. D. (1991). Ellagic acid, an anticarcinogen in fruits, especially in strawberries. *HortScience*, *26*, 10−14.

Minoggio, M., Bramati, L., Simonetti, P., Gardana, C., Lemoli, L., Santangelo, E., Mauri, P. L., Spigno, P., Soressi, G. P. & Pietta, P. G. (2002). Polyphenol pattern and antioxidant activity of different tomato lines and cultivars. *Annals of Nutrition and Metabolism*, *47*, 64–69.

Milivojević, J., Nikolić, M. & Đurović, D. (2008). The influence of growing system on cropping potential of strawberry cultivar 'Clery' grown in plastic tunnel. *VI International Strawberry Symposium*, *842*, pp. 115–118.

Milivojević, J., Nikolić, M. & Bogdanović-Pristov, J. (2010). Fizičko-hemijska i antioksidativna svojstva sorti i samoniklih vrsta rodova Fragaria i Rubus. *Voćarstvo*, *44*, 169/170, 55–64.

Milivojević, J., Maksimović, V., Nikolić, M., Bogdanović, J., Maletić, R. & Milatović, D. (2011). Chemical and antioxidant properties of cultivated and wild Fragaria and Rubus berries. *Journal of Food Quality*, *34*, 1−9.

Milić, B., Đilas, S., Čanadanović-Brunet, J. & Sakač, M. (2000). *Biljni polifenoli.* Matica srpska, Novi Sad.

Montero, T. M., Molla, E. M., Esteban, R. M. & López-Andréu, F. J. (1996). Quality attributes of strawberry during ripening. *Scientia Horticulturae*, *65*, 239–250.

Moore, U., Karp, K. & Põldma, P. (2004). Effect of mulching and fertilization on the quality of strawberries. *Agricultural and Food Science*, *13*, 256–267.

Mosel, H. D. & Herrmann, K. (1974). Phenolics in fruits. The phenolics of blackberries and raspberries and their changes during development and ripenessof the fruits. *Zeitschrift fur Lebensmittel Untersuchung und Forschung*, *154*, 324–327.

Moyer, R., Hummer, K., Wrolstad, R. E. & Finn, C. (2002). Antioxidant compounds in diverse Ribes and Rubus germplasm. *Acta Horticulturae*, *585*, 501–505.

Murillo, E., Britton, G. B. & Durant, A. A. (2012). Antioxidant activity and polyphenol content in cultivated and wild edible fruits grown in Panama. *Journal of Pharmacy and Bioallied Sciences*, *4*, 313–317.

Narayana, K. R., Reddy, M. S., Chaluvadi, M. R. & Krishna, D. R. (2001). Bioflavonoids classification, pharmacological, biochemical effects and therapeutic potential. *Indian Journal of Pharmacology*, *33*, 2–16.

Nieman, D. C., Henson, D. A., McAnulty, S. R., McAnulty, L., Swick, N. S., Utter, A. C., Vinci, D. M., Opiela, S. J. & Morrow, J. D. (2002). Influence of vitamin C supplementation on oxidative and immune changes after an ultramarathon. *Journal of Applied Physiology*, *92*, 1970−1977.

Nikkhah, E., Khaiamy, M., Heidary, R. & Azar, A. S. (2010). The effect of ascorbic acid and H_2O_2 treatment on the stability of anthocyanin pigments in berries. *Turkish Journal of Biology*, *34*, 47–53.

Nile, S. H. & Park, S. W. (2014). Edible berries: Bioactive components and their effect on human health, *Nutrition*, *30*, 134–144.

Olsson, M. E., Gustavsson, K. E., Andersson, S., Nilsson, A. & Duan, R. D. (2004). Inhibition of cancer cell proliferation *in vitro* by fruit and berry extracts and correlations with antioxidant levels. *Journal of agricultural and food chemistry*, *52*, 7264−7271.

Omaye, S. T. & Zhang P. (1998). Phytochemical interactions: ß-carotene, tocopherol and ascorbic acid. In: '*Phytochemicals*'. Bidlack, W.R., Omaye, S.T., Meskin, M.S., Jahner, D. (eds.), Technomic Publishing Co., Lancaster, USA, pp. 53–75.

Pantelidis, G. E., Vasilakakis, G., Mangaris, G. A. & Diamantidis G. (2007). Antioxidant capacity, phenol, anthocyanin and ascorbic acid contents in raspberries, blackberries, red currants, gooseberries and Cornelian cherries. *Food Chemistry*, *102*, 777–783.

Parr, A. J. & Bolwell, J. P. (2002). Phenols in the plant and in man. The potential for possible nutritional enchacement of the diet by modifying the phenols content or profile. *Journal of the Science of Food and Agriculture*, *80*, 985–1012.

Pellegrini, N., Searfini, M., Colombi, B., Del Rio, D., Salvator,e S., Bianchi, M. & Brighenti, F. (2003). Total antioxidant capacity of plant, foods, beverages and oils consumed in Italy assessed by three different *in vitro* assays. *Journal Nutrition*, *133*, 2812−2819.

Pereira, D. M., Valentão, P. & Andrade, P. (2009). Phenolics: from chemistry to biology. *Molecules*, *14*, 2202–2211.

Perez-Tello, G. O., Silva-Espinoza, B. A., Vargas-Arispuro, I., Briceño-Torres, B. O. & Martinez-Tellez, M. A. (2001). Effect of temperature on enzymatic and physiological factors related to chilling injuri in carambola fruit (Averrhoa carambola L.). *Biochemical and Biophysical Research Communication*, *287*, 846–851.

Pešaković, M. & Milivojević, J. (2014). Comparative study of bio- and chemical fertilization in strawberry production. In: *Fertilizers: components, uses in agriculture and environmental impacts*. Fernández-Luqueño F., López-Valdez F., (eds.). Nova Science Publishers, Inc., New York, USA, pp. 127−154.

Pešaković, M., Milenković, S., Đukić, D., Mandić, L., Karaklajić-Stajić, Ž., Tomić, J. & Miletić N. (2016). Phenolic composition and antioxidant capacity of integrated and conventionally grown strawberry (Fragaria x ananassa Duch.). *Horticultural Science*, *42*, 4, 200−206.

Pineli, L. D. L. D. O., Moretti, C. L., dos Santos, M. S., Campos, A. B., Brasileiro, A. V., Córdova, A. C. & Chiarello, M. D. (2011). Antioxidants and other chemical and physical characteristics of two strawberry cultivars at different ripeness stages. *Journal of Food Composition and Analysis*, *24*, 11−16.

Pinto, M. D. S., Lajolo, F. M. & Genovese, M. I. (2007). Bioactive compounds and antioxidant capacity of strawberry jams. *Plant foods for human nutrition*, *62*, 127−131.

Prior, R. L. & Cao, G. (2000). Analysis of botanicals and dietary supplements for antioxidant capacity: a review. *Journal of AOAC International*, *83*, 950−956.

Prochzáková, D., Boušova, I. & Wilhelmova, N. (2011). Antioxidant and prooxidant properties of flavonoids. *Fitoterapia*, *82*, 513–523.

Proteggente, A. R., Pannala, A. S., Paganga, G., Van Buren, L., Wagner, E., Wiseman S., Van de Put, F., Dacombe, C. & Rice-Evans, C. A. (2002). The antioxidant activity of regularly consumed fruit and vegetables reflects their phenolic and vitamin C composition. *Free Radical Research*, *36*, 217–233.

Puupponen-Pimia, R., Nohynek, L., Meier, C., Kähkönen, M., Heinonen, M. & Hopia, A. (2001). Antimicrobial properties of phenolic compounds from berries. *Journal of Applied Microbiology*, *90*, 4, 494–507.

Rana, R. K. (2001). *Studies on the influence of nitrogen fixers and plant bioregulators on growth, yield and quality of strawberry cv. Chandler*. Doctoral dissertation. University of Horticulture and Forestry.

Reverberi, M., Picardo, M., Ricelli, A., Camera, E., Fanelli, C. & Fabbri, A. A. (2001). Oxidative stress, growth factor production and budding in potato tubers during cold storage. *Free Radical Research*, *35*, 833–841.

Reyes-Carmona, J., Yousef, G. G., Martinez-Peniche, R. A. & Lila, M. A. (2005). Antioxidant capacity of fruit extracts of blackberry (Rubus sp.) produced in different climatic regions. *Journal of Food Science*, *70*, 497–503.

Rice-Evans, C., Miller, N. & Paganga, G. (1997). Antioxidant properties of phenolic compounds. *Trends in plant science*, *2*, 152−159.

Robards, K. & Antolovich, M. (1997). Analytical chemistry of fruit bioflavonoids. *A review. Analyst*, *122*, 11−34.

Robards, K., Prenzler, P. D., Tucker, G., Swatsitang, P. & Glover, W. (1999). Phenolic compounds and their role in oxidative processes in fruits. *Food Chemistry*, *66*, 401–436.

Robbins, R. J. (2003). Phenolic acids in foods: an overview of analytical methodology. *Journal of Agricultural and Food Chemistry*, *51*, 10, 2866–2887.

Romero-Rodriguez, M. A., Vazquez Oderiz, M. L., Lopez Hernandez, J. & Simal Lozano, J. S. (1992). Determination of vitamin C and organic acids in various fruits by HPLC. *Journal of Chromatographic Science*, *30*, 11, 433–437.

Rutz, J. K., Voss, G. B. & Zambiazi, R. C. (2012). Influence of the degree of maturation on the bioactive compounds in blackberry (Rubus spp.) cv. Tupi. *Food and Nutrition Sciences*, *3*, 10, 1453–1460.

Sacks, F. M. & Kass, E. H. (1988). Low blood pressure in vegetarians: effects of specific foods and nutrients. *The American journal of clinical nutrition*, *48*, 795−800.

Sariburun, E., Sahin, S., Demir, C. & Türkben Uylaşer, V. (2010). Phenolic content and antioxidant activity of raspberry and blackberry cultivars. *Journal of Food Science*, *75*, 4, 328–335.

Satue-Garcia, T., Heinonen, M. & Frankel, E. N. (1997). Anthocyanins as antioxidants on human low-density lipoprotein and lecithin–liposome systems. *Journal of Agricultural and Food Chemistry*, *45*, 3362–3367.

Scalzo, J., Politi, A., Pellegrini, N., Mezzetti, B. & Battino, M. (2005). Plant genotype affects total antioxidant capacity and phenolic contents in fruit. *Nutrition*, *21*, 207–213.

Schouten, S., Hopmans, E. C., Forster, A., van Breugel, Y., Kuypers, M. M. & Damsté, J. S. S. (2003). Extremely high sea-surface temperatures at low latitudes during the middle Cretaceous as revealed by archaeal membrane lipids. *Geology*, *31*, 12, 1069–1072.

Schuster, B. & Herrmann, K. (1985). Hydroxybenzoic and hydroxycinnamic acid derivatives in soft fruits. *Phytochemistry*, *24*, 2761–2764.

Seeram, N. P. (2009). Bioactive polyphenols from foods and dietary supplements: challenges and opportunity. In: *Herbs: Challenges in Chemistry and Biology*. Ho, C.T., Wang, M., Sang S. (eds.), Oxford University Press, New York, USA, pp. 5308–5312.

Sellapan, S., Akoh, C. C. & Krewer, G. (2002). Phenolic compounds and antioxidant capacity of Georgia-Grown blueberries and blackberries. *Journal of Agriculture and Food Chemistry*, *50*, 2432–2438.

Shahidi, F. & Naczk, M. (1995). *Food phenolics*. Technomic Pub. Co.

Shahidi, F. & Naczk, M. (2003). *Phenolics in food and nutraceuticals*. Taylor & Francis Group, Florida, USA.

Siriwoharn, T. & Wrolstad, R. E. (2004). Polyphenolic composition of Marion and Evergreen blackberries. *Journal of Food Science*, *69*, 4, 233–240.

Spanos, G. A. & Wrolstad, R. E. (1992). Phenolics of apple, pear, and white grape juices and their changes with processing and storage. *Journal of Agriculture and Food Chemistry*, *40*, 1478–1487.

Stajčić, S., Tepić, A., Djilas, S., Šumić, Z., Čanadanović-Brunet, J., Ćetković, G. & Vulić, J. (2012). Chemical composition and antioxidant activity of berry. *Acta Periodica Technologica*, *43*, 93–105.

Stanisavljević, M. (1999). New small fruit cultivars from Čačak: 1. The new blackberry (Rubus sp.) cultivar 'Čačanska Bestrna'. *Acta Horticulturae*, *505*, 291−296.

Stintzing, F. C., Stintzing, A. S., Carele, R., Frei, B. & Wrolstad, R. E. (2002). Color and antioxidant properties of cyanidin-based anthocyanin pigments. *Journal of Agricultural and Food Chemistry*, *50*, 6172–6181.

Stöhr, H. & Herrmann, K. (1975). The phenolics of fruits. VI. The phenolics of currants, gooseberries and blueberries. Changes in phenolic acids and catechins during development of black currants. *Zeitschrift fur Lebensmittel-Untersuchung und Forschung*, *159*, 31–37.

Strack, D., Wray, V. (1993). The anthocyanins. In: '*The flavonoids: Advances in research since 1986*'. Harborne, J.B. (ed.), Chapman & Hall, London, UK, pp. 1–22.

Strack, D. (1997). Phenolic metabolism. In: '*Plant biochemistry*'. Dey, P.M., Harborne, J.B. (eds.), Academic Press, London, UK, pp. 387–389.

Sumbul, S., Ahmad, M. A., Asif, M. & Akhtar, M. (2011). Role of phenolic compounds in peptic ulcer: An overview. *Journal of Pharmacy and Bioallied Sciences*, *3*, 3, 361–367.

Szajdek, A. & Borowska, E. J. (2008). Bioactive compounds and health-promoting properties of berry fruits. *Plant Foods for Human Nutrition*, *63*, 147–156.

Tomás-Barberán, F. A. & Clifford, M. N. (2000). Dietary hydroxybenzoic acid derivatives–nature, occurrence and dietary burden. *Journal of the Science of Food and Agriculture*, *80*, 1024−1032.

Tomic, J., Milivojevic, J. & Pesakovic, M. (2015). The response to bacterial inoculation is cultivar-related in strawberries. *Turkish Journal of Agriculture and Forestry*, *39*, 332–341.

Tomić, J. (2016). *Effect of microbiological and mineral fertilizers on biological and production characteristics of strawberry cultivars* (*Fragaria ananassa* Duch.). University of Belgrade, Faculty of Agriculture, 1–190.

Truax, B., Gagnon, D., Lambert, F. & Chevrier, N. (1994). Nitrate assimilation of raspberry and pin cherry in a recent clearcut. *Canadian Journal of Botany*, *72*, 1343−1348.

Tulipani, S., Mezzetti, B., Capocasa, F., Bompadre S., Beekwilder, J., Ric de Vos, C. H., Capanoglu, E., Bovy, A. & Battino, M. (2008). Antioxidants, phenolic compounds, and nutritional quality of different strawberry genotypes. *Journal of Agricultural and Food Chemistry*, *56*, 696−704.

Tulipani, S., Alvarez-Suarez, J. M., Busco, F., Bompadre, S., Quiles, J. L., Mezzetti, B. & Battino, M. (2011a). Strawberry consumption improves plasma antioxidant status and erythrocyte resistance to oxidative haemolysis in humans. *Food chemistry*, *128*, 180−186.

Tulipani, S., Marzban, G., Herndl, A., Laimer, M., Mezzetti, B. & Battino, M. (2011b). Influence of environmental and genetic factors on health-related compounds in strawberry. *Food Chemistry*, *124*, 906−913.

Umar, I., Wali, V. K., Kher, R. & Jamwal, M. (2009). Effect of Fym, Urea and Azotobacter on Growth, Yield and Quality of Strawberry cv. Chandler. *Notule Botanicae Horti Agrobotanici Cluj Napoca*, *37*, 139–143

Vasco, C. (2009). *Phenolic compounds in Ecuadorian fruits.* Doctoral thesis. Swedish University of Agricultural Sciences, Uppsalla.

Vauzour, D. (2012). Dietary polyphenols as modulators of brain functions: biological actions and molecular mechanisms underpinning their beneficial effects. *Oxidative medicine and cellular longevity*, *2012*, 1−16.

Veberič, R., Slatnar, A., Jakopič, J., Štampar, F. & Mikulič Petkovšek, M. (2012). Primarnii sekundarni metaboliti u voću. *Zbornik radova i apstrakata 14. kongresa voćara i vinogradara Srbije sa međunarodnim učešćem*, Vrnjačka Banja, Srbija, 55–62.

Veberic, R., Stampar, F., Schmitzer, V., Cunja, V., Zupan, A., Koron, D. & Mikulic-Petkovsek, M. (2014). Changes in the contents of anthocyanins and other compounds in blackberry fruits due to freezing and long-term frozen storage. *Journal of Agricultural and Food Chemistry*, *62*, 6926–6935.

Veberič, R., Slatnar, A., Bizjak, J., Stampar, F. & Mikulic-Petkovsek, M. (2015). Anthocyanin composition of different wild and cultivated berry species. *LWT - Food Science and Technology*, *60*, 509−517.

Voća, S., Duralija, B., Družić, J., Skenderović-Babojelić, M., Dobričević, N. & Čmelik, Z. (2006). Influence of cultivation systems on physical and chemical composition of strawberry fruits cv. Elsanta. *Agriculturae Conspectus Scientificus*, *71*, 171–174.

Wang, H., Cao, G. & Prior, R. L. (1996). Total antioxidant capacity of fruits. *Journal of Agricultural and Food Chemistry*, *44*, 701−705.

Wang, H., Cao, G. & Prior, R. L. (1997). Oxygen radical absorbing capacity of anthocyanins. *Journal of Agricultural and Food Chemistry*, *45*, 304−309.

Wang, S. Y. & Camp, M. J. (2000). Temperatures after bloom affect plant growth and fruit quality of strawberry. *Scientia Horticulturae*, *85*, 183−199.

Wang, S. Y. & Jiao, H. (2000). Scavenging capacity of berry crops on superoxide radicals, hydrogen peroxide, hydroxyl radicals and singlet oxygen. *Journal of Agricultural and Food Chemistry, 48*, 5677–5684.

Wang, S. Y. & Lin, H. S. (2000). Antioxidant activity in fruits and leaves of blackberry, raspberry and strawberry varies with cultivar and developmental stage. *Journal of Agricultural and Food Chemistry, 48*, 140–146.

Wang, S. Y. & Zheng, W. (2001). Effect of plant growth temperature on antioxidant capacity in strawberry. *Journal of Agricultural and Food Chemistry, 49*, 4977–4982.

Wald, B., Galensa, R., Herrmann, K., Grotjahn, L. & Wray, V. (1986). Quercetin 3-O-[3-hydroksi-3-methylglutaroyl)-ß-galactoside] from blackberries. *Phytochemistry, 25*, 12, 2904–2905.

Williner, M. R., Pirovani, M. E. & Güemes, D. R. (2003). Ellagic acid content in strawberries of different cultivars and ripening stages. *Journal of the Science of Food and Agriculture, 83*, 842–845.

Wrolstad, R. E. (1976). *Color and pigment analyses in fruit products*. Corvallis, Ore.: Oregon Agric Expt Station Bull, 624.

Wrolstad, R. E., Skrede, G., Lea, P. & Enersen, G. (1990). Influence of sugar on anthocyanin pigment stability in frozen strawberries. *Journal of Food Science, 55*, 1064–1072.

Yoshiki, Y., Okubo, K. & Igarashi, K. (1995). Chemiluminescence of anthocyanins in the presence of acetaldehyde and tert-butyl hydroperoxide. *Journal of bioluminescence and chemiluminescence, 10*, 335−338.

Zhao, Y. (2007). *Berry fruit. Value-added products for health promotion*. CRS Press, Taylor and Francis Group, Boca Raton, Florida, USA.

Zheng, W. & Wang, S. Y. (2003). Oxygen radical absorbing capacity of phenolics in blueberries, cranberries, chokeberries, and lingonberries. *Journal of Agricultural and Food Chemistry, 51*, 502−509.

In: Phenolic Compounds
Editors: P. Oliver and A. Villem
ISBN: 978-1-53612-881-9

Chapter 3

RECOVERY OF PHENOLIC COMPOUNDS FROM OLIVE MILL WASTES: A REVIEW

M. Almeida[1], R. C. Martins[1], M. D. Víctor-Ortega[2,*], L. Gando-Ferreira[1] and R. M. Quinta-Ferreira[1]

[1]CIEPQPF - Chemical Process Engineering and Forest Products Research Centre, Department of Chemical Engineering, Faculty of Sciences and Technology, University of Coimbra, Coimbra, Portugal

[2]Department of Chemical Engineering, University of Granada, Granada, Spain

ABSTRACT

Olive oil production is an important industry especially in the Mediterranean Sea area where almost 97% of the worldwide olive oil is produced. These industries are generally characterized by small mills spread over large areas. Moreover, olive oil extraction generates large

---

* Corresponding Author: Tel: +34958241581; Fax: +34958248992; E-mail: mdvictor@ugr.es.

amounts of liquid and solid wastes which are characterized by high toxicity and seasonal character. The most generalized management strategy for these wastes is dewatering and soil spreading. However, the increasing concern regarding the impact of this procedure over the soil and aquifers quality is pushing towards the development of more environmentally friendly approaches. One important feature of olive mill wastes is the presence of phenolic compounds. In fact, olives are rich in those compounds and a high fraction is lost during the extraction process. Only about 2% of phenolic compounds are retained in the olive oil. Thus up to 98% are incorporated in solid and liquid wastes. Phenolic compounds are related with the bactericide and phytotoxic character of these wastes which difficult their biological treatment.

Phenolic compounds present a high variety of biologic activities such as antioxidant, cardioprotector, anti-inflammatory and chemopreventive. Thus these substances present a high added-value and olive mill wastes seem to be an interesting source. In this context, phenolic compounds recovery will integrate added-value compounds obtained through renewable sources with waste management. Several recovery strategies are presented in literature involving extraction, filtration and adsorption that may be integrated with advanced oxidation treatments to minimize the environmental impact of the remaining wastes. In this context, this chapter aims to make a short overview on the works published regarding phenolic compounds recovery from olive mill liquid wastes.

Keywords: environment, olive mill wastes, phenolic compounds, recovery

1. INTRODUCTION

The increasing demands of population are leading to the lowering of the quantity and quality of fresh water. Moreover, the discharge of untreated or poorly treated wastewaters, namely those coming from olive mills, is leading to the contamination of the natural water sources. This is an important threat for both ecosystems and human health. The increasing social concern regarding environmental issues is imposing political commitment for developing more restrictive laws for wastewater discharge regulation.

A particular problematic case of industrial pollution is related with olive oil production. Especially during the extraction processes large amounts of olive mill wastewaters (OMW) are generated. Those streams are characterized by high organic charge, seasonality and low biodegradability. Moreover, are known by their bactericide and phytotoxic nature. Those characteristics are related with their high load in phenolic compounds that are hardly removed by biological processes. This means that those traditional wastewater treatments are not suitable for the management of OMW. Thus, in the last decades efforts were putted in action for the development of suitable OMW treatments (Martins et al., 2010; Lucas and Peres, 2009).

On the other hand, phenolic compounds, especially those of low molecular weight, at the right doses have health benefits. In fact, they present interesting characteristics such as: antioxidant, anti-inflammatory and cardio protector capacities. Thus, those substances may be used in several industrial sectors such as food, cosmetics and pharmaceutical (Victor-Ortega et al., 2016). Thus, phenolic substances (namely for example, hydroxytyrosol), are considered as compounds of high added value with high market prices. Considering the high concentration on those substances in OMW, attention is starting to focus on the possibility of recovering such compounds from OMW. This way raw materials would be available at low costs; besides, there would be environmental benefits due to the removal of those refractory pollutants from the wastewater.

Bearing this in mind, this chapter aims to give an overview on the methodologies based on membranes and adsorption processes applied for the recovery of phenolic compounds from OMW.

2. Olive Oil Production

Almost 97% of olive oil production is concentrated in the Mediterranean region. Nowadays, most of olive oil is produced through one of three systems: traditional press, three phases centrifugal decantation and two phases centrifugal decantation.

The traditional press system is based on olive oil removal from the olive through pressure followed by solid-liquid and liquid-liquid separations. This process produces up to 0.6 m^3 of OMW per 1000 kg of olives (Azbar et al., 2004).

In the three phases decantation system, horizontal centrifuges are used for liquid-solid separation. The separation is aided by the addition of water. Thus three phases are produced: oil, water and solids. The high disadvantage of this system is the high amount of water used which will lead to high loads of OMW. Thus per 1000 kg of olives processed, 1.0 to 1.2 m^3 of OMW are generated. More recently, this process evolved to a low water consuming system denominated by two-phases decantation. In this technology, lower amounts of water are used leading to lower loads of OMW (0.12 m^3 per 1000 kg of olive). Nevertheless, in this case, pollutants are mainly transferred to a moisture solid waste.

OMW composition is variable and depends on the edaphoclimatic conditions of olive production, characteristics and duration of olives storage and, mostly, from the technology applied for olive oil extraction (Zbakh and El Abbassi, 2012; Cabrera et al., 1996). Generally, OMW is acidic, with high conductivity, brown coloration and with several dissolved and suspended compounds. Typical composition (wt.) is 83% to 94% in water, 4 – 16% in organic compounds (fats, sugars, nitrogenous substances, organic acids, polyalcohols, pectins, tannins and phenolic compounds) and 0.4 - 2.5% of minerals. Thus, these streams are characterized by high contaminant features with high values of chemical and biochemical oxygen demand (Vlyssides et al., 1998; De Marco et al., 2007; Davies et al., 2004).

The presence of phenolic compounds (more than 50 were already identified) gives to OMW the brown color as well as the bactericide and phytotoxic character (Cabrera et al., 1996; Zbakh and El Abbassi, 2012).

3. OMW Valorization

There are several studies aiming to give an environmental solution to OMW by increasing its commercial potential. According to Ramos-

Cormenzana et al. (1995), there are several valorization strategies that may be applied to OMW:

1) Biogas production;
2) Natural antioxidants (phenolic compounds) valorization for food, pharmaceutical and cosmetic industry application;
3) Protein production for animals' food production;
4) Substrate for algae growth, polysaccharide production or ethanol production.

Olives are known by their high load in phenolic compounds. Only 2% are incorporated in the olive oil. Thus up to 98% are disposed in the OMW (Galanakis et al., 2010). In this context, OMW are rich in a high diversity of phenolic compounds. Among them hydroxytyrosol, oleurepein, tyrosol, caffeic acid, vannilic acid and p-coumaric acid can be referred. All these compounds present high biologic activity and antioxidant capacity. Thus, due to their high added value, their recovery is an interesting strategy for OMW management.

3.1. Membrane Processes

Membrane separation processes are well stablished technologies with applications in food and beverages production as well as wastewater treatment (Galanakis et al., 2010). These systems may be applied for the recovery and concentration of phenolic compounds in OMW and may involve microfiltration (MF), ultrafiltration (UF), nanofiltration (NF) or reverse osmosis (RO). Table 1 summarizes works dealing with the application of membrane based processes for phenolic compounds recovery from OMW.

Generally speaking before the application of UF, NF or RO, a pre-treatment is used for the removal of suspended solids that may lead to membrane fouling.

Table 1. Overview of publications dealing with the application of membrane based processes for phenolic compounds recovery from OMW

Technology	Operating Conditions	Concluding Remarks	Reference
UF	OMW from 3 phases extraction. Pre-treatment: MF (0.2 μm, T = 25°C and ΔP = 0.5 bar); UF: regenerated cellulose membranes (4, 5, and 10 kDa), T = 25°C and ΔP = 2 or 5 bar.	Phenolic compounds recovered in permeate (up to 48.3%). The pre-treatment reduced fouling.	Cassano et al., 2011
UF and NF	OMW from 3 phases extraction. Pre-treatment: MF; I. UF (0.02 μm, T = 18°C and ΔP = 0.43 bar); II. UF (1000 Da, T = 30°C and ΔP = 9 bar); III. NF (90 Da, T = 29°C and ΔP = 5 e 9 bar).	I. Used to remove suspended solids. 26% of phenolic compounds rejected; II. Recovery of up to 31.8% of phenols; III. Leads to a permeate rich in phenolic compounds (93%).	Cassano et al., 2013
UF, NF and RO	Pilot-scale cross flow batch filtration: I.UF (100 nm and ΔP = 1 bar); II. NF (95% of $MgSO_4$ rejection, ΔP = 20 bar); III. RO (99% of NaCl rejection, ΔP = 40 bar). Permeate of each step is used as feed in the following one.	I. UF: 62% of the phenolic compounds are in the concentrate; II. NF: Removes up to 90% of the compounds with molecular weight (MW) > 468 g/mol. 28% of phenols remain in the concentrate; III. RO: 9% of phenols remain in the concentrate (MW < 500 g/mol).	Zagklis and Paraskeva, 2014

Technology	Operating Conditions	Concluding Remarks	Reference
UF, NF and RO	Pilot-scale cross flow batch filtration: I.UF (100 nm, T = 15 – 50ºC and ΔP = 1 – 2.5 bar); II. NF (200 Da, ΔP = 10 - 30 bar); III. RO (100 Da, ΔP =20 - 40 bar). Permeate of each step is used as feed in the following one.	I. UF: the increase on the transmembrane pressure increases phenols removal; II. NF: pressure increase diminishes the phenolic content in the permeate; III. RO: similar behavior to NF.	Paraskeva et al., 2007
MF, NF	I. MF (200 nm, ΔP = 0.72 bar) II. NF (578 Da, ΔP = 8 bar)	I. MF: recovers 78% of phenolic compounds in the permeate; Hydroxytyrosol was the main phenolic compound (54%); II. 5% of phenolic compounds rejection; Hydroxytyrosol was the main phenolic compound (56%).	Garcia-Castello et al., 2010
UF	10 kDa, pH 3, 5 and 9, ΔP = 0.5 - 2 bar	Pressure does not significantly influences phenolic compounds removal (50%). pH rules both pollutants and membrane charge. At pH 9 phenolic compounds show the highest rejection (60%).	Martins et al., 2015

Table 2. Overview of publications dealing with the application of ionic exchange processes for phenolic compounds recovery from water

Resins	Operating Conditions	Concluding Remarks	Reference
Amberlyst A26 (anionic strong); Amberlite IRA-67 (anionic weak)	Simulated phenol solution; [Phenol] = 5 – 100 mg/L; [Resin] = 3.5 g/L; T = 25°C	Equilibrium in 1h; Adsorption capacity increase with phenol concentration; pH increase favors phenolate production increasing adsorption. Maximum adsorption capacity was 148.6 mg/L for A26 and 76.3 mg/L for 76.3 mg/L. Desorption using NaOH (4%) lead to almost total phenol recovery.	Victor-Ortega et al., 2016
Amberlite XAD7 (polar weak), XAD16 (no-polar), IRA 96 (polar). Isolute ENV+ (no-polar)	OMW from a three phases extraction method; T = 25°C, 180 rpm, 1 h	ENV+ lead to the best adsorption results (84% of phenolic compounds). Desorption: acidified ethanol recovered 96% and 82% of phenolic compounds in XAD16 and ENV+, respectively.	Bertin et al., 2011
Amberlite XAD7 (polar weak), XAD16 (no-polar), IRA 96 (polar). Isolute ENV+ (no-polar)	Simulated solution of 10 phenolic compounds; T = 25°C, 180 rpm, 1 h	IRA96 and ENV+ lead to 89% and 64.6% of adsorption, respectively. Desorption: acidified ethanol lead to the best results.	Ferri et al., 2011

Resins	Operating Conditions	Concluding Remarks	Reference
Amberlite XAD-4 (no-polar). Amberlite XAD-4-I and XAD-4-II (obtained by chemical modification of XAD-4)	Simulated phenolic solution; Batch tests: T = 298, 308 and 318 K; t = 8h; 220 rpm.	Chemical modification of the resin increased up to 50% of surface area and pore volume. Adsorption increased up to 20%.	Li et al., 2013
Amberlite IRA-420 (anionic strong)	Phenol solution. Batch tests: T = 298 K and t = 48 h.	Adsorption mechanism highly influenced by pH: high pH ion exchange prevails while for low pH adsorption is the key step.	Carmona et al., 2006
Macronet MN200; Dowex XZ (anionic strong); Aurix 100 (anionic weak)	Phenol solution. Batch tests: T = 21°C; pH 3 - 11	MN200 with best results and higher affinity for molecular phenol. Dowex XZ and Auris 100 with better results for alkaline conditions due to phenol dissociation into phenolate ion. Desorption: MN200 90% of phenol recovered using a methanol/water solution.	Caetano et al., 2009
Amberlite XAD4 (no-polar). Amberlite XAD16N. Amberlite XAD7HP	OMW concentrate from RO. Batch tests with different solid/liquid (S/L) ratios. Desorption using several solvents (water, ethanol and acetone).	XAD4 with the highest phenols removal (up to 80% for S/L 120). Desorption with highest efficiency using acetone (80% of phenolic compounds recovery leading to a product with 378 $g_{phenolic\ compounds}$/L.	Zagklis et al., 2015

Table 2. (Continued)

Resins	Operating Conditions	Concluding Remarks	Reference
Amberlite: XAD4, XAD7HP, XAD761, XAD16, FPX66. Lewatit: AF5, AF6, AF7. Monoplus M800, K6387, VPOC1600, VPOC1064, MDPH. Monoplus SP112. Activated carbon: GAC, CAL-1	Simulated effluent (hydroxytyrosol, tyrosol, oleuropein, luteolin, caffeic acid). Real OMW pre-treated by UF.	Luteolin was highly adsorbed by all resins. Oleuropein: highly adsorbed by XAD4, XAD16, FPX66 and VPOC1064. AF5 shows higher selectivity for hydroxytyrosol and tyrosol (almost 100%). pH of OMW is a key factor for the process.	Kaleh and GeiBen, 2016

Afterwards, OMW is subjected to UF to clarify permeate without the removal of a significant fraction of phenolic compounds. That removal is function of the membrane nature, its cut-off as well as of the characteristics of OMW (Paraskeva et al., 2007). Polymeric membranes are the most usually applied; however, fouling is a problem. Fouling in ceramic membranes seems to be less problematic.

NF is usually able to concentrate phenolic compounds leading to permeate with low organic matter (di Lecce et al., 2014). RO is the most efficient process obtaining a cleaner permeate. However, the pressure required may be economically not viable for industrial applications.

3.2. Adsorption/Ionic Exchange

The use of non-ionic resins for the removal of hydrocarbons from water is widely spread. Nevertheless, the application of anionic resins seems more efficient for phenolic compounds recovery from water. Table 2 shows some works dealing with the application of ion exchange resins for phenolic substances removal from water.

Ion-exchange resins seem to be more selective since the process encompasses not only adsorption mechanism but also ion exchange. Ion exchange efficiency is generally improved for alkaline conditions due to the phenolic compounds ionization into phenolates (Carmona et al., 2006). Strong ionic resins lead to better removal efficiencies especially (up to 94%) at high pH (Victor-Ortega et al., 2016). However, most of the data available is for simulated solutions and further studies are required for actual OMW.

CONCLUSION

OMW constitute a serious environmental threat. Their characteristics do not allow the application of the traditional biological treatments for wastewater management. This is especially due to the presence of phenolic compounds that are known by their bactericide and phytotoxic features. However, these substances may be applied in food, pharmaceutical and cosmetic industry with high added value. This way, their recovery from OMW is interesting both economically and environmentally. The use of membrane separation technologies coupled with adsorption/ion exchange seems to be an interesting option. Although the promising results obtained with synthetic solutions, further studies are required using actual OMW.

ACKNOWLEDGMENTS

Fundação para a Ciência e Tecnologia is gratefully acknowledged by the financial support under the contract FCT investigator 2014 Program (IF/00215/2014) with financing from European Social Fund and the Human Potential Operational Program (R.C. Martins).

REFERENCES

Azbar, N., Bayram, A., Filibeli, A., Muezzinoglu, A., Sengul, F, Ozer, A. (2004). A review of waste management options in olive oil production. *Crit. Rev. Environ.*, 34, 209-247.

Bertin, L., Ferri, F., Scoma, A., Marchetti, L., Fava, F. (2011) Recovery of high added value natural polyphenols from actual olive mill wastewater through solid phase extraction. *Chem. Eng. J.*, 171, 1287-1293.

Cabrera, F., López, R., Martinez-Bordiú, A., Dupuy de Lome, E., Murillo, J. (1996). Land treatment of olive mill wastewater. *Int. Biodeterior. Biodegradation.,* 38, 215-225.

Caetano, M., Valderrama, C., Farran, A., Cortina, J. (2009) Phenol removal from aqueous solution by adsorption and ion exchange mechanisms onto polymeric resins. *J. Colloid Interface Sci.*, 338, 402-409.

Carmona, M., De Lucas, A., Valverde, J., Velasco, B., Rodríguez, J. (2006) Combined adsorption and ion exchange equilibrium of phenol on Amberlite IRA-420. *Chem. Eng. J.*, 117, 155-160.

Cassano, A., Conidi, C., Drioli, E. (2011) Comparison of the performance of UF membranes in olive mill wastewaters treatment. *Water Res.*, 45, 3197-3204.

Cassano, A. (2013) Fractionation of olive mill wastewaters by membrane separation techniques. *J. Hazard. Mater.*, 248, 185-193.

Davies, L., Vilhena, A., Novais, J., Martins-Dias, S. (2004) Olive mill wastewater characteristics: modelling and statistical analysis. *Grasas y Aceites.*, 55, 233-241.

De Marco, E., Savarese, M., Paduano, A., Sacchi, R. (2007) Characterization and fractionation of phenolic compounds extracted from olive oil mill wastewaters. *Food Chem.*, 104, 858-867.

Di Lecce, G., Cassano, A., Bendini, A., Conidi, C., Giorno, L., Toschi, T. (2014) Characterization of olive mill wastewater fractions treatment by integrated membrane process. *J. Sci. Food Agric.*, 94, 2935-2942.

Ferri, F., Bertin, L., Scoma, A., Marchetti, L., Fava, F. (2011) Recovery of low molecular weight phenols through solid-phase extraction. *Chem. Eng. J.*, 166, 994-1001.

Galanakis, C., Tornberg, E., Gekas, V. (2010). Clarification of high-added value products from olive mill wastewater. *J. Food Eng.*, 99, 190-197.

Garcia-Castello, E., Cassano, A., Criscuoli, A., Conidi, C., Drioli, E. (2010) Recovery and concentration of polyphenols from olive mill wastewaters by integrated membrane system. *Water Res.*, 44, 3883-3892.

Kaleh, Z., GeiBen, S. (2016) Selective isolation of valuable biophenols from olive wastewater. *Biochem. Farmacol.*, 4, 373-384.

Li, C., Xu, M., Xiucheng, S., Deng, S. (2013) Chemical modification of Amberlite XAD-4 by carbonyl groups for phenol adsorption from water. *Chem. Eng. J.*, 229, 20-26.

Lucas, M. S., Peres, J. (2009) Removal of COD from olive mill wastewater by Fenton's reagent: kinetic study. *J. Hazard. Mater.*, 168, 1253-1259.

Martins, R. C., Silva, A., Castro-Silva, S., Garção-Nunes, P., Quinta-Ferreira, R. (2010). Adopting strategies to improve the efficiency of ozonation in the real scale treatment of olive oil mill wastewaters. *Environ. Technol.,* 31(13), 1459-1469.

Martins, R. C., Ferreira, A., Gando-Ferreira, L., Quinta-Ferreira, R. (2015) Ozonation and ultrafiltration for the treatment of olive mill wastewaters: Effect of key operating conditions and integration schemes. *Env. Sci. Poll. Res.*, 22, 15587-15597.

Paraskeva, C., Papadakis, V., Tsarouchi, E., Kanellopoulou, D., Koutsoukos, P. (2007) Membrane processing for olive mill wastewater fractionation. *Desalination*, 213, 218-229.

Ramos-Cormenzana, A., Monteoliva-Sanchez, M., Lopez, M. (1995). Bioremediation of alpechin. *Int.Biodeterior. Biodegradation.*, 35, 249-268.

Victor-Ortega, M., Ochando-Pulido, J., Martínez-Ferez, A. (2016). Performance and modeling of continuous ion exchange processes for phenols recovery from olive mill wastewater. *Process Saf. Environ.*, 100, 242-251.

Vlyssides, A., Loizidou, M., Gimouhopouls, K., Zorpas, A. (1998) Olive oil processing wastes production and their characteristics in relation to olive oil extraction methods. *Fresen. Environ. Bull.*, 12, 603-611.

Zagklis, D., Paraskeva, C. (2014) Membrane filtration of agro-industrial wastewaters and isolation of organic compounds with high added value. *Water Sci. Technol.*, 69, 202-207.

Zagklis, D., Vavouraki, A., Kornaros, M., Paraskeva, C. (2015) Purification of olive mill wastewater phenols through membrane filtration and resin adsorption/desorption. *J. Hazard. Mater.*, 285, 69-76.

Zbakh, H., El Abbassi, A. (2012) Potential use of olive mill wastewater in the preparation of functional beverages: A review. *J. Funct. Foods.*, 4, 53-65.

In: Phenolic Compounds
Editors: P. Oliver and A. Villem
ISBN: 978-1-53612-881-9

Chapter 4

PHENOLIC ANTIOXIDANTS ON CALIXARENE AND CALIXRESORCINOL SCAFFOLDS

***Sergey Bukharov*[1], *Gulnara Nugumanova*[1] *and Alexander Burilov*[2]**

[1]Kazan National Research Technological University of the Ministry of Education and Science of Russia, Russia

[2]A. E. Arbuzov Institute of Organic and Physical Chemistry of Kazan Scientific Center of Russian Academy of Sciences, Russia

ABSTRACT

The literature data on antiradical and antioxidizing properties of calix[n]arenes and calix[n]resorcinols have been summarized. The dependence of these properties of the composition and structure of macrocycles and methods for the modification of calix[n]arenes and calix[n]resorcinols by hindered phenolic fragments has been considered. The effect of "preorganization" of antioxidant fragments attached to calix[n]arene and calix[n]resorcinol scaffolds and their intramolecular synergy on antioxidant activity has been discussed.

INTRODUCTION

In the recent two decades, 3D structures, in particular, calix[n]arenes and calix[n]resorcinol macrocycles, which represent unique "building blocks" for the synthesis of supramolecular systems, are thoroughly investigated in organic chemistry. On the other hand, these compounds are platforms for the location of functional groups possessing various types of activity.

Due to "preorganization" effect, functional groups on calix[n]arene matrix are capable of cooperative effect, which would result in the manifold increase in the effectiveness of a particular valuable effect. As an example, it is known that groups, which exhibit or not wear complexing properties, may form an effective complexing system in calix[4]arene due to spatial preorganization [1, 2]. There are examples of a drastic increase in biological activity of compounds due to the formation of calix[4]arene macrocycle. For example, calix[4]arene **1** exhibits high antibacterial activity at the level of hexamidine with respect to *Escherichia coli*, *Staphylococcus aureus*, *Enterococcus faecalis*, *Pseudomonas aeruginosa*, and others. In this case, compound **2**, which can be considered monomeric equivalent of calix[4]arene **1**, does not display activity with respect to mentioned causative pathogens [3].

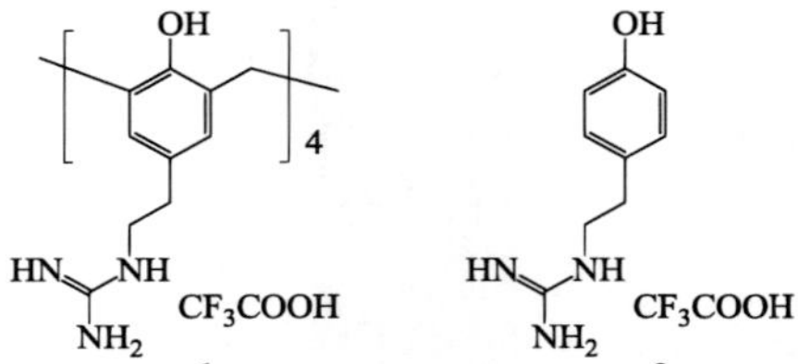

1 **2**

One of the promising directions of the practical application of calix[n]arenes and calix[n]resorcinols is their employment as polyfunctional stabilizers-antioxidants. Recently, thermal and photostabilizing action of calix[n]arenes, calix[n]resorcinols, and some of their derivative in organic materials are known. They manifested themselves as effective autoxidation inhibitors towards polyolefins and plastics, radioprotectors, and modifiers for thermosetting resins [4-9].

An interest in calix[n]arene and calix[n]resorcinol scaffolds as a molecular basis for the development of new effective inhibitors of radical-chain oxidation is caused by the following factors:

- Calix[n]arenes and calix[n]resorcinols as polyphenolic compounds possess antioxidant properties.
- These platforms open broad opportunities for their functionalization and, consequently, for the design of various "hybrid" structures. They provide the compounds with various combinations of polar and nonpolar properties (different hydrophilic–lipophilic balance), which are soluble in and compatible with the media of different polarity and polymers. This fact is significant both for stabilization of polymers and the design of drug preparations.
- Structural preorganization of calix[n]arene and calix[n]resorcinol platforms offers an opportunity for the cooperative effect of the attached additional antioxidant fragments, which in fact may provide their higher antioxidizing action.

The present review is devoted to the systematization of the literature data on antioxidant properties of calix[n]arenes and calix[n]resorcinols, consideration of the dependence of these properties on the composition and structure of macrocycles, and methods of modification of calix[n]arenes and calix[n]resorcinols by hindered phenols.

1. Polymer Stabilizers Based on Calix[n]arenes

An early work regarding calix[n]arenes as antioxidants for polymers is represented by the patent of L.P.J. Burton [4], where phenol-phosphite adducts prepared by the reaction of phosphorus trichloride with calixarenes **3** and **4** in the presence of lithium hydride are described.

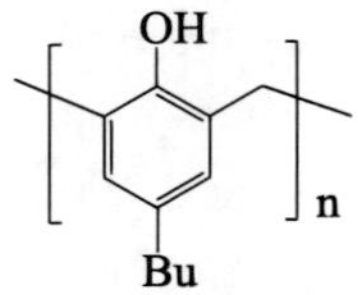

3 (n=4), **4** (n=6)

The structure of the products is not specified by the author and the composition was characterized only by elemental analysis data. It was shown that phenol-phosphates adducts both individually and in the mixture with dilauryl thiodipropionate provides an effective protection of polypropylene under thermal-oxidative ageing.

Seiffarth and Goermar et al. [7] showed that *para-n*-butylcalix[4]arene **3** provides a significantly larger induction period during initiated oxidation of tetralin as compared to 2,6-di-*tert*-butyl-4-methylphenol (BHT).

The authors from [10, 11] suggested the mechanism of stabilizing action of calix[n]arenes, which involves the cleavage of kinetic oxidation chains, which is intrinsic for phenolic antioxidants:

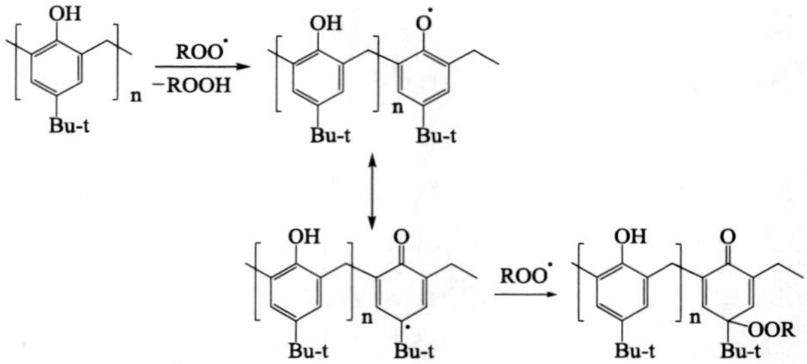

High antioxidation activity of calix[n]arenes can be presumably rationalized by relative stability of formed phenoxyl radicals, which is caused by the possibility of additional delocalization of unpaired electron along macrocyclic system:

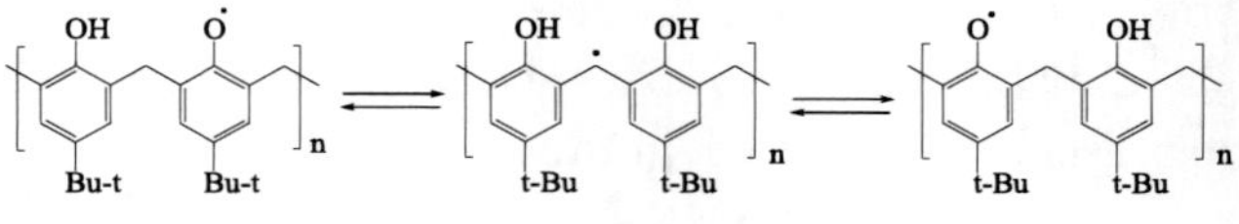

This effect may be confirmed by the fact that phenolic compounds bearing hydrogen atoms at α-carbon atom of *ortho*-substituent possess increased antioxidizing activity, which is caused by regeneration of phenol from phenoxy radical [12-15]:

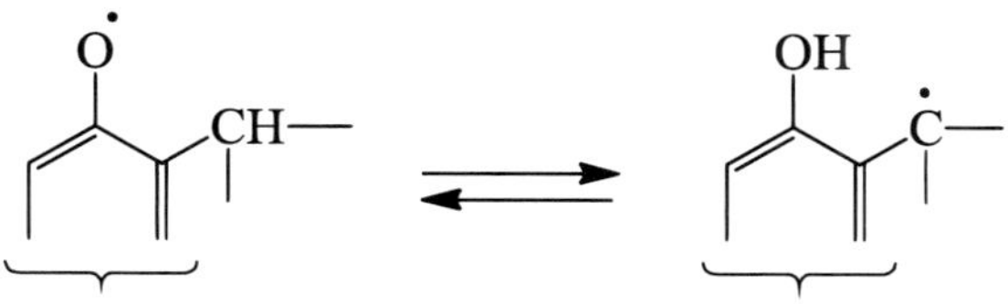

Delocalization of unpaired electron of phenoxyl radical along macrocyclic system may also be assisted by the presence of cyclic hydrogen bond between hydroxyl groups of calix[n]arenes [16].

In authors' opinion from [17], another factor, which is favorable for the application of calix[n]arenes as antioxidants of organic polymers, is their high thermal stability. In this case, it should be noted that there is no direct relationship between thermal stability of calix[n]arenes and their antioxidant activity. For example, the authors from [18] compared the kinetics of thermal destruction of calix[6] and calix[8]arenes **6** and **7** with their antioxidant activity with respect to polypropylene.

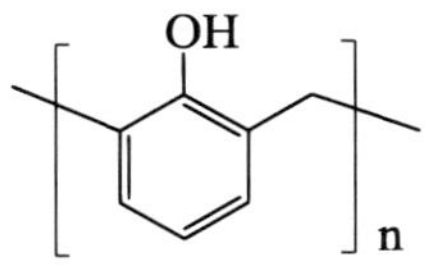

6 (n=6), **7** (n=8)

EPR, ^{13}C NMR, and chromatomass spectrometry methods showed that the thermal destruction of these calix[n]arenes proceeds with the cleavage of macrocycle and the formation of open-chain polyphenolic radicals. According to the data obtained by authors, calix[8]arene **7** possesses lower thermal stability (lower activation energy of destruction) and higher antioxidant activity during thermal destruction of polypropylene as compared to calix[6]arene **6**.

A tendency to the increase in stabilizing activity of calix[n]arenes with an increase in the number of phenolic fragments in molecules is also observed during thermal oxidation of high-density polyethylene (HDPE) [10]. The values of induction periods of oxidation of stabilized HDPE specimens, which were determined by chemiluminescent method, demonstrate higher effectiveness of para-*tert*-butylcalix[6]arene **4** as compared to para-*tert*-butylcalix[4]arene **3**.

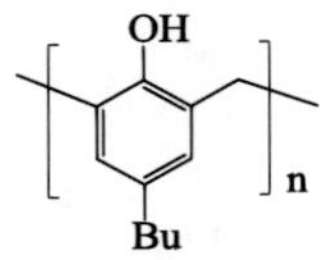

3 (n=4), **4** (n=6), **8** (n=8)

In contrast, there is an opposite result under the conditions of thermal oxidizing ageing of low-density polyethylene (LDPE); calix[4]arene **3** containing four phenolic fragments in molecule was more effective. An identical series of protector activity of calix[n]arenes, namely, calix[4]arene **3** > calix[6]arene **4** > calix[8]arene **8**, was observed during thermal oxidizing ageing of LDPE [11].

Higher effectiveness of calix[n]arenes in LDPE than HDPE was associated by the authors with structural features of these polymers, more specifically, crystallinity and branching. Calix[n]arenes accept more mobile polymer peroxide radicals of linear structure, which is intrinsic for LDPE, presumably more effectively.

Antioxidizing stabilization of organic compounds and polymers by calix[n]arenes was also mentioned in the patents from [5, 19, 20].

An important area of practical application of hindered phenols is the protection of polymers against radioactive degradation. Polypropylene is the most frequently employed polymer in the fabrication of single-use medicinal products. Its radiation and post-radiation degradation can be minimized by the introduction of appropriate hindered phenols [21].

In [22], the protective action of para-*tert*-butylcalix[4]arene **3** during oxidative radiolysis of polypropylene is reported presumably for the first time. The authors from [9, 23] compared the protective effect of para-*tert*-

butylcalix[n]arenes **3**, **4**, **8** during γ-radiation of polypropylene. The protective effect was determined according to the data of mechanical tests of irradiated polypropylene by tensiometry. It was determined that the protector activity of calixarene grows in the following order: calix[8]arene **8** < calix[6]arene **4** = 2,6-di-*tert*-butyl-4-methylphenol (BHT) < calix[4]arene **3**.

Later, the authors from [24] investigated the effect of para-*tert*-butylcalix[n]arenes **3, 4** on the main kinetic parameters of oxidation of isotactic polypropylene at accelerated thermal and radiation ageing. It was determined that calix[4]arene **3** provides the larger induction period of thermal oxidation of polypropylene (τ=260 min) as compared to calix[6]arene **4** and high-performance industrial antioxidant represented by 2,6-di-*tert*-butyl-4-octadecylpropionylphenol (Irganox 1076) (τ=150 and 210 min, respectively). However, during radiation ageing calix[4]arene, calix[6]arene, as well as Irganox 1076 did not show high stabilizing activity according to parameters mentioned above, which in authors' opinion is caused by their low intrinsic γ-radiation stability. Stabilizing effect of para-*tert*-butylcalix[n]arenes during thermal degradation of polyproline was also studied in [25].

In [6], acylated calix[n]arenes of general formula **9** as inhibitors of thermal oxidation destruction of polyolefins; polystyrene; ABS plastics; butadiene, isoprene, and natural rubbers; polyesters; and lubricants were patented.

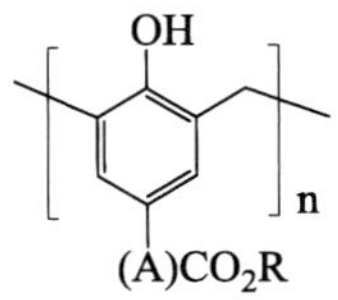

9

A is a direct bond or is alkylene of 1 to 2 carbon atoms; R, when A is a direct bond, is alkyl of 4 to 30 carbon atoms, cycloalkyl of 5 to 6 carbon atoms, phenyl or phenyl substituted by alkyl of 1 to 18 carbon atoms; R, when A is alkylene, is alkyl of 1 to 30 carbon atoms, cycloalkyl of 5 to 6 carbon atoms, phenyl or phenyl substituted by alkyl of 1 to 18 carbon atoms; n is 1-7.

Calix[n]arenes **9** exhibit a remarkable stabilizing action under thermal oxidation ageing of polypropylene and ABS plastics, while their effectiveness exhibits a superadditive growth in the presence of typical radical-free hydroperoxide degradative agents represented by dilauryl thiodipropionate and distearyl thiodipropionate. In [26], calixarenes of formula **10**, which were prepared by the reaction of calix[n]arene glycidyl esters with amines, alcohols, and carboxylic acids, were suggested for the protection of polymers, oils, and low-molecular organic compounds from ageing under the action of heat, oxygen, and light.

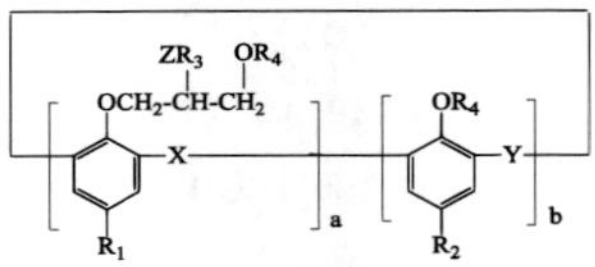

10

X=Y= -$(CR_5R_6)_c$; R_5=R_6= hydrogen, alkyl, cycloalkyl, aryl, arylalkyl; c = 1-3; X=Y= -CH_2OCH_2-, -CO-; Z= -O-, -S-, -NR_7; R_7= hydrogen, alkyl, cycloalkyl, aryl, arylalkyl or R_5-CO- group; R_1=R_2=hydrogen, substituted or unsubstituted aryl, W-$(CR_5R_6)_d$; W=R_5-O-CO-, R_5R_6N-CO-, R_5-CO-, halogen, R_5R_6N-, aryl, cycloalkyl; d = 1-30;

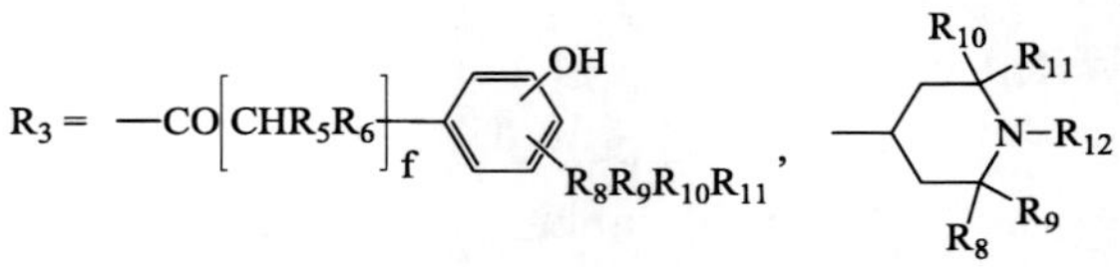

where R_8,R_9,R_{10},R_{11}= hydrogen, alkyl, cycloalkyl, aryl, arylalkyl, OH group, f=0-30; R_{12}= hydrogen, alkyl, cycloalkyl, aryl, hydroxyl, R_5-CO-, R_5-O-; R_3=9-20 C alkyl; R_4= hydrogen, alkyl, cycloalkyl, R_5-CO-; a=1-12; b=0-11; a+b=n=2-12

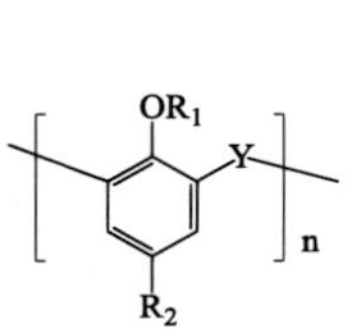

11

n=2-12; Y= -C(R_3R_4)-, -C(R_3R_4)-O-C(R_3R_4)-, -C=O; R_1,R_2,R_3= H, alkyl, cycloalkyl, arylalkyl, aryl.

In order to stabilize various polymers (polyolefins, ABS plastics, polyamides, polyesters, and polyurethanes), low-molecular organic compounds, and oils, molecular complexes, which consist of the mixture of cyclic polynuclear methylenephenolic compounds of general formula **11** are suggested [8].

With the aim to create synergetic effects, the authors recommend to use their combinations with other known additives possessing various functionalities, such as hindered phenols, 2-hydroxybenzophenone and benzothiazole derivatives, organic phosphites, alkyldithiopropionates, and amines.

As an effective thermal stabilizing additive for polyolefins, polystyrene, poly(vinyl acetate), poly(vinyl chloride), poly(methyl methacrylate), ABS plastics, and polyurethanes, the compositions containing heat stabilizers such as metal salts of fatty acids (zinc, calcium, and barium stearates), polyols (pentaerythritol, sorbitol, 1,3-butanediol, and others), epoxy compounds (epoxy soya bean oil, dioctylepoxyhexahydrophthalate), as well as calix[n]arenes of general formula **12** are suggested [27].

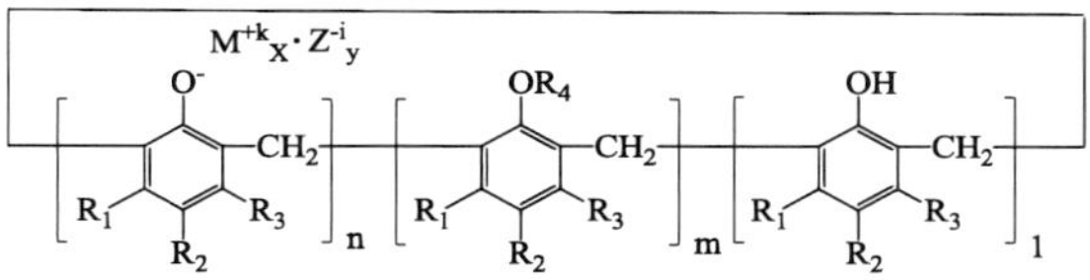

12

where R_1,R_2,R_3 – hydrogen atom, saturated or unsaturated alkyl group, aryl group, alkoxyl group, halogen atom, nitro group, acyl group, carboxyl group, sulfonic acid group, amino group; R_4 – saturated or unsaturated alkyl group, aryl group, acyl group; n = m = 0-10, n + m + 1 = 4-10; M^{+k} – metal ion, NH_4^+ ion, organic cation, +k – the valence number of the ion, k = 1-6; Z^{-i} – anion, -i – the valence number of the anion, i = 1-6; x = 0-10; y = 0-10.

To avoid thermal oxidation of poly(vinyl chloride), *para*-isopropenylcalix[8]arene **13** and its octaacetate **14** are also suggested [28]. DTA method showed that the introduction of compounds **13** and **14** to PVC compositions leads to the increase in the temperature corresponding to the maximum rate of dehydrochlorination and temperatures corresponding to 5–50% weight loss of polymer. The period before the onset of dehydrochlorination of PVC specimens 4 times as large as the analogous value in the case of PVC, which does not contain calix[8]arenes.

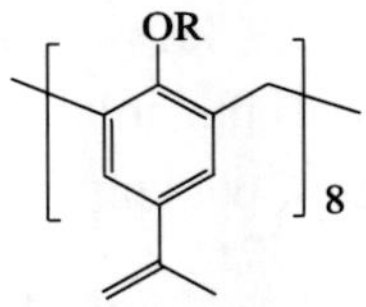

R = H (**13**), $CH_3C(O)$ (**14**)

Thermal characteristics of the specimens of PVC compositions stabilized by compounds **13** and **14**, are almost identical to the characteristics of the specimen stabilized by industrial Naftomix DWX 200A lead-containing additive.

In [29], hydroxycinnamic acid derivatives on calix[4]arene platform **15** and **16**, which are of significant interest as antioxidants and antiradical agents, were described.

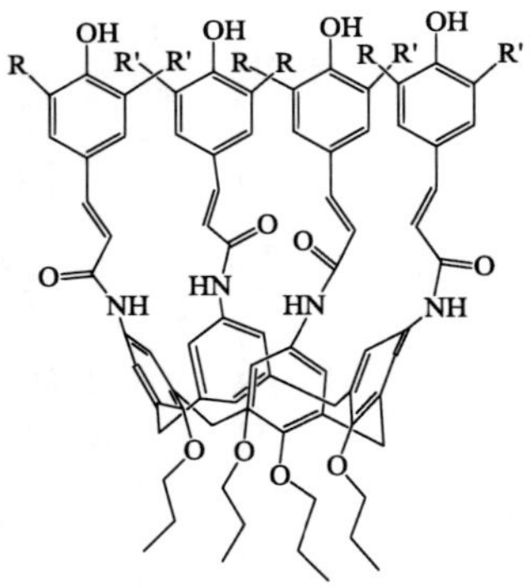

R=OH, R'=H (**15**)
R=R'=OCH_3 (**16**)

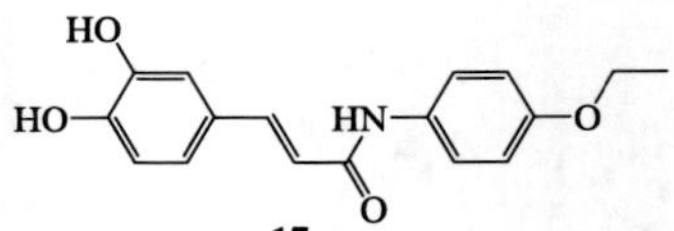

17

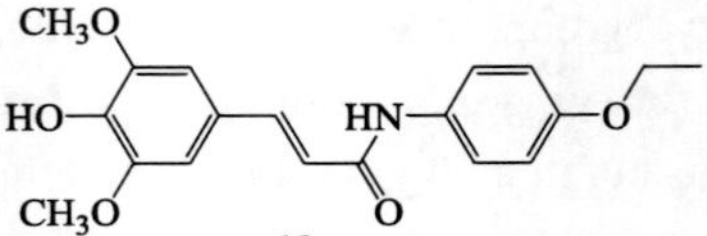

18

Compounds **15** and **16** are intrinsic for high rate constants of the interaction with 2,2-diphenyl-1-picrylhydrazyl radical (k1 > 1.5•104) and stoichiometric coefficients n (7.7 and 2.7 for compounds **15** and **16**, respectively). They exceed remarkably analogous values for compounds **17** and **18**, which are considered model units of conjugates **15** and **16**.

It was also demonstrated that the antioxidant activity of calix[4]arenes **15** and **16**, which was determined during 2,2′-azobisisobutyronitrile-initiated oxidation of linoleic acid, exceeds that of *para*-phenetidine derivatives **17** and **18**. These results indicate an important role of preorganization of phenolic groups on calixarene platform to exhibit their cooperative effect, which leads to the increase in antioxidant activity of stabilizer.

Phenolic groups of various types are also present in calix[4]arene **19**, which was described in [30]:

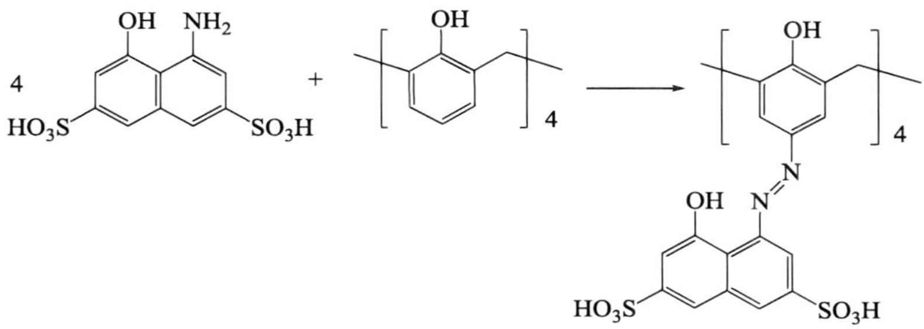

19

Compound **19** exhibited high antioxidant properties in the "β-carotene – linoleic acid" model system and high antiradical activity in the reaction with 1,1-diphenyl-2-picrylhydrazyl.

Several groups of researchers discovered independently the photostabilizing activity of calix[n]arenes in polyolefins. In [7, 31], it was determined that the activity of calix[n]arenes as photostabilizers depends on the size of macrocycle and types of substituents at *para*-position of phenolic rings. The authors rationalize this dependence by the effect of mentioned parameters on the melting point and compatibility of calix[n]arenes with polyolefins. The introduction of tertiary carbon atom to *para*-position of phenolic ring provides an increase in photostabilizing activity of calix[n]arene, while primary and secondary carbon atoms increase antioxidant activity. Calix[n]arenes possessing high melting are distributed weakly in polyolefins and possess low photostabilizing action. Calix[n]arenes, which are highly compatible with polyolefins, possess the same photostabilizing action as industrial photostabilizers based on 2-hydroxybenzophenone. The authors associate this fact with the oxidation of calix[n]arenes under UV radiation and alkylhydroperoxides, which are formed during polyolefin processing.

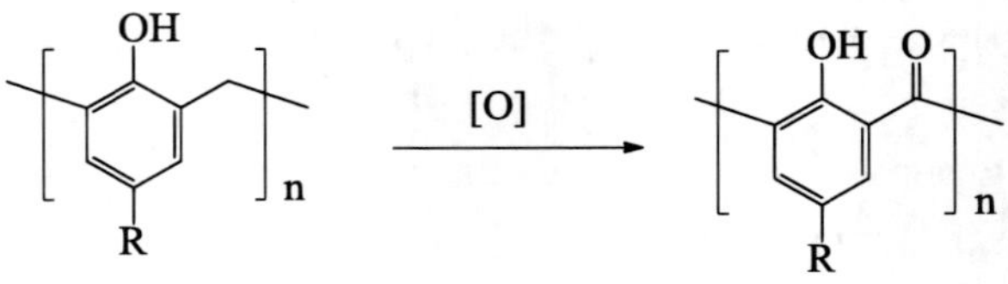

The products of oxidation of calix[n]arenes contain carboxylic groups in bridging fragments, which provides their photostabilizing activity at the degree of 2-hydroxybenzophenone.

High-performance heat stabilizers of polymers possessing photostabilizing action are represented by *tert*-butylcalix[4]arene **20**, one of the methods for preparation of which was described in [32]:

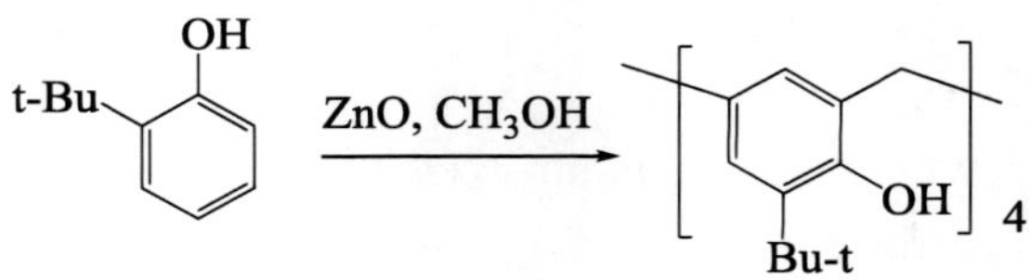

20

2. Polymer Stabilizers Based on Calix[n]resorcinols

Calix[4]resorcinols also possess antioxidant properties. The investigations carried out by Ehrhardt D. et al. [33-36] showed that tetramethylcalix[4]resorcinol **21** may be used as antioxidant in colored and carbon black reinforced rubbers based on natural, butadiene-nitrile, butadiene-styrene, and chloroprene rubbers.

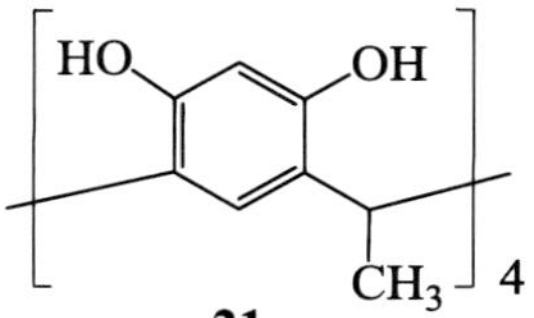

21

According to the effectiveness of protective action in thermal oxidation processes, calix[4]resorcinol **21** exceeds hindered alkylated mono- and bisphenols. Compound **21** to some extent may replace amine stabilizers in rubber mix, such as phenyl-β-naphthylamine (Neozon D) and polymerized 2,2,4-trimethyl-1,2-dihydroquinoline (Antioxidant TMQ). Calix[4]resorcinol **21** provides anti-fatigue protection of rubbers based on the mixture of natural and butadiene-styrene rubbers increasing their fatigue cracking resistance by a factor of 1.5–2.5. The effectiveness of calix[4]resorcinol is higher than that of 2,2′-methylene-bis-(4-methyl-6-*tert*-butylphenol) by 10 – 40%.

Tetramethylcalix[4]resorcinol **21** combined with amine stabilizers also provide antiozonant action, which results in the possible decrease in the content of amine stabilizers in rubber compounds. Results of investigation of atmosphere stability of resin based on natural and butadiene-styrene rubbers showed that the combination of calix[4]resorcinol **21** with N-isopropyl-N'-phenyl-p-phenylenediamine (Antioxidant 4010 NA) is more effective than the composition of Antioxidant 4010 NA with Neozon D.

In [37], the synthesis of calix[4]resorcinol **22** was carried out:

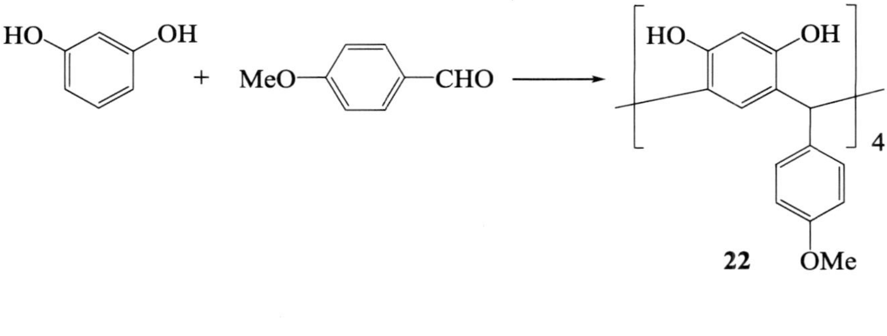

22

In the model reaction with 1,1-diphenyl-2-picrylhydrazyl, this compound showed moderate antiradical activity.

In the patent from [38], high-performance photostabilizers of organic materials based on substituted calix[n]arenes **23** and calix[n]resorcinols **24** were suggested:

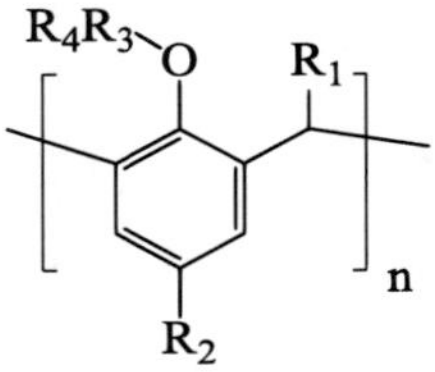

23

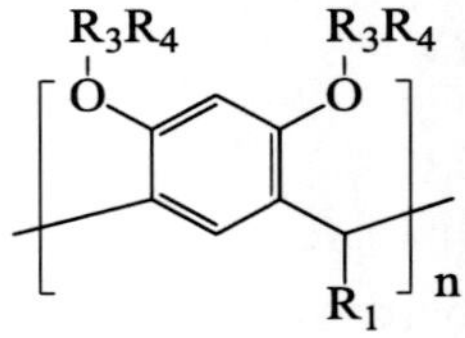

24

n = 1-8; R_1 – H, C_1-C_{18} alkyl, C_3-C_{10} cycloalkyl, C_1-C_4 alkyl, trifluoromethyl, C_1-C_4 alkoxy, halogen, cyan, carboxy, nitro, amino, C_1-C_4 alkylamino, C_1-C_4 dialkylamino, acyl, acyloxy; R_2 – C_1-C_{18} alkyl; R_3 = $-CH_2C(O)-$, -C(O)-, -C(O)-C(O)-, $-CH_2CHR_1$, $-(CH_2CH_2O)_x-$, x=1-8, R_4 – fragment, containing hindered amine, for example:

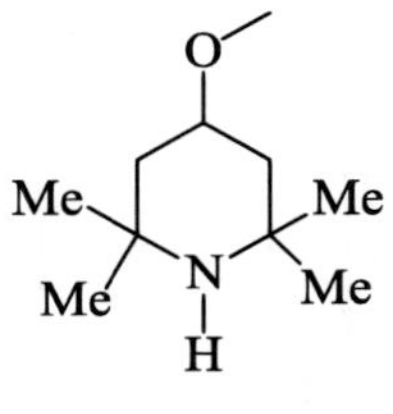

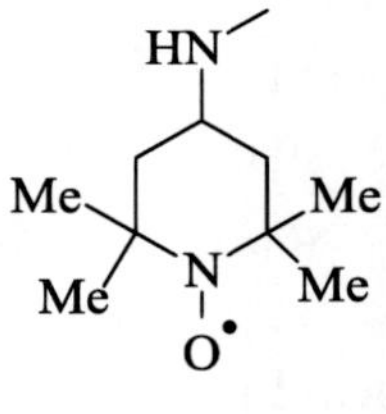

Tetranitroxides **26a-c** [39], which were prepared by aminomethylation of calix[4]resorcinols **21**, **25a,b** with 4-amino-2,2,6,6-tetramethylpiperidinyloxy radical (4-amino-TEMPO) and formaldehyde, exhibit high antioxidant and antiradical activity.

In the model reaction with 1,1-diphenyl-2-picrylhydrazyl radical, compounds **21**, **26a-c** exhibit similar reactivity. However, the values of relative antioxidizing effectiveness prove that tetranitroxides of calix[4]resorcinols **26a-c** are 10 times more effective in the inhibition of 2,2′-azobis(2-amidopropane)dihydrochloride-initiated oxidation of linoleic acid than calix[4]resorcinol **21**. On the whole, calix[4]resorcinols **21**, **25a,b** и **26a-c** are ~100 times more effective as radical traps than 4-amino-TEMPO and resorcinol. This can be related to the possible synergetic effect of two types of radical peroxide traps in compounds **26a-c**, as well as delocalization of spin, which is formed upon the cleavage of hydrogen atom from calix[4]resorcinols **21**, **25a,b**, along the resorcinol macrocycle.

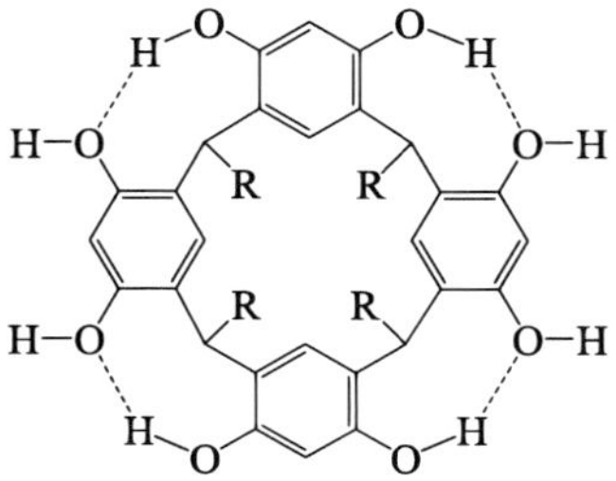

R=CH_3 (**21**), C_2H_5 (**25a**), C_7H_{15} (**25b**)

R_1 =

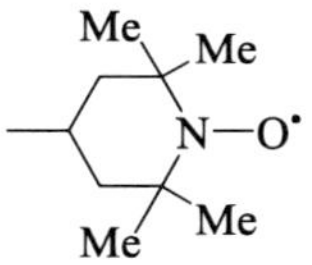

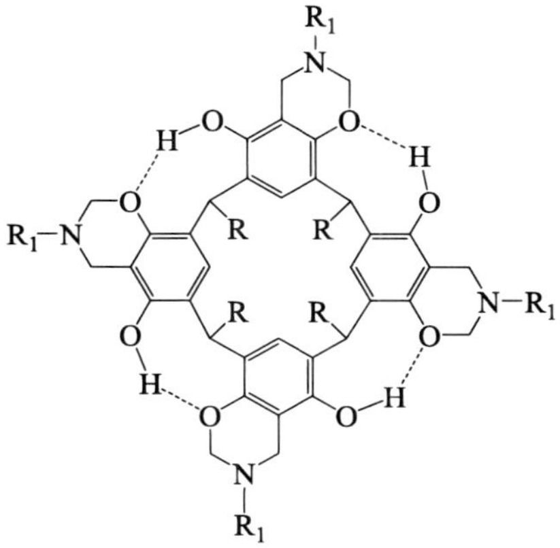

R=CH_3 (**26a**), C_2H_5 (**26b**), C_7H_{15} (**26c**)

High stoichiometric inhibition coefficients of tetranitroxide calix[4]resorcinols **26a-c** are associated by the authors with the possibility of multiple regeneration of TEMPO-fragments in the transformation cycle of nitroxyl radicals in the reactions with alkyl and peroxide.

Thus, the investigation of antioxidant activity of calix[n]arenes and calix[n]resorcinols showed the possibility of the design of high-performance polymer stabilizers on their basis. Modification of these macrocycles with hindered phenolic fragments is a promising method for the increase in their antioxidant activity. Such modification provides the synthesis of the compounds containing phenolic groups, which differ in reactivity, which may be important in the reactions of inhibited oxidation. The macrocycles modified by this route may exhibit the effect of intramolecular synergy according to the mechanism of regeneration of more active peroxide radical trap represented by phenol or resorcinol hydroxyl with the formation of more stable hindered phenolic radical:

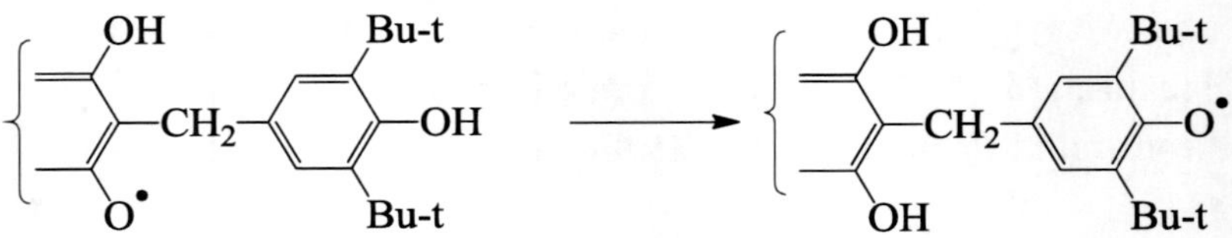

3. CALIX[N]ARENES MODIFIED BY HINDERED PHENOLIC FRAGMENTS

The first hindered phenolic derivatives of calix[4]arene were presumably described in [40]:

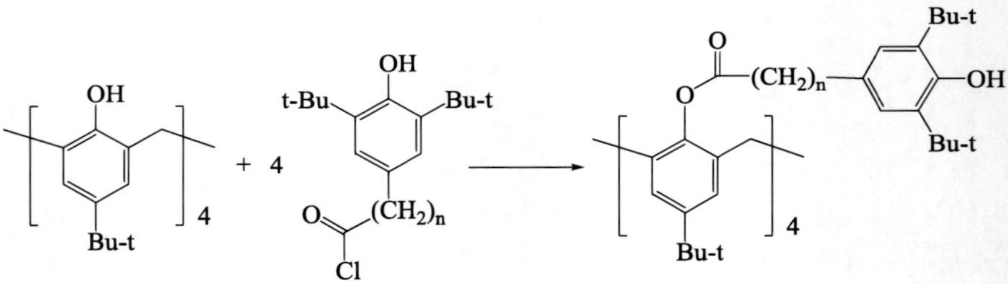

3 n = 0 (**a**), 2 (**b**) **27a,b**

Based on ^{1}H NMR spectroscopy data, the authors determined that calix[4]arene **27a** exists in *partial cone* conformation. Compounds **27a** and **27b** displayed higher antioxidant activity during initiated oxidation of tetralin and autoxidation of LDPE than initial para-*tert*-butylcalix[4]arene **3**. Modified calix[4]arene **27b** exhibited the highest antioxidant activity, which exceeds analogous values of both initial calix[4]arene and phenolic stabilizers **28** and **29**. This fact also indicates a significant influence of the preorganization of phenolic fragments described above on antioxidant activity.

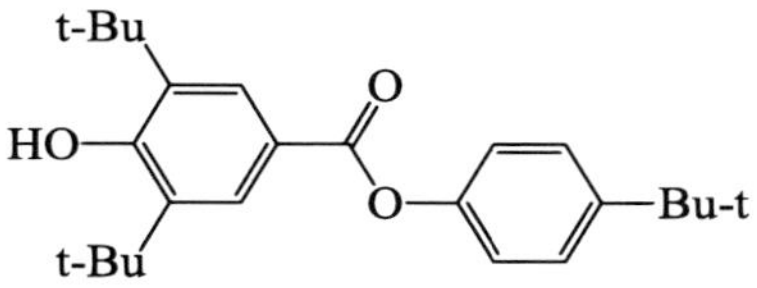

28

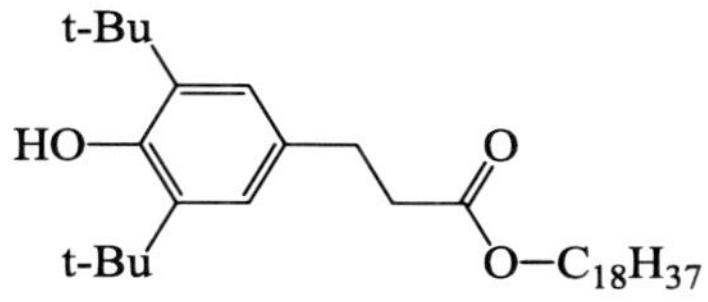

29

Synthesis of phosphorylated calix[4]arenes **30a,b,c** containing hindered phenolic fragments on the lower rim of macrocycle is described in [41]:

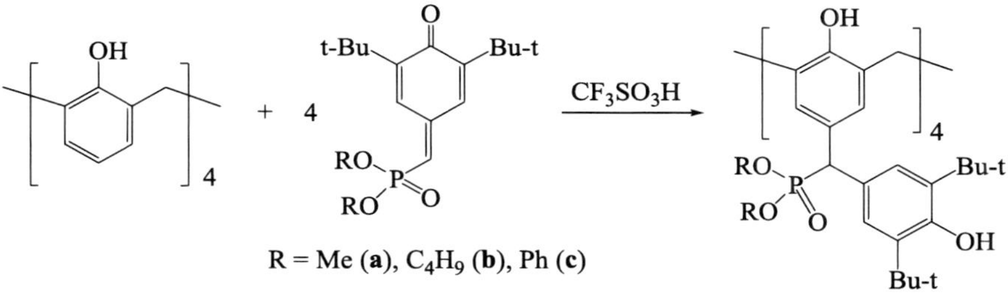

R = Me (**a**), C_4H_9 (**b**), Ph (**c**)

30a,b,c

Thiacalix[n]arenes containing phenolic hydroxyl groups and sulfide bridges represent a particular interest for the synthesis of polyfunctional antioxidants. These compounds may simultaneously act as peroxide radical traps and radical-free destruction agents of hydroperoxides and exhibit the effect of intramolecular synergy of antioxidant activity. In [42], antioxidant activity of *para*-tert-butylcalix[4]arene and *para*-tert-butylthiacalix[4]arene during azobisisobutyronitrile-initiated oxidation of methyl oleate were compared. The discovered lower antioxidant activity of thiacalix[4]arene is associated by the authors with the ability of sulfides to form thienyl radicals, which may participate in the oxidation chain propagation step. It should be noted that experimental conditions used in [42], more specifically, initiated oxidation, do not allow the evaluation of the contribution of radical-free destruction of hydroperoxides to the antioxidant activity of this compound. The total antioxidant activity of *para*-tert-butylthiacalix[4]arene, which is caused both by its ability to form kinetic chains of oxidation in the reactions with peroxide radicals and destruct hydroperoxides without radical, can be evaluated during autoxidation conditions of experiments.

Additional functionalization of thiacalix[n]arenes by hydrazone groups extends remarkably their antioxidant opportunities. It is known that hydrazones possess thermal and photostabilizing features and may act as peroxide radical traps [43, 44] The complexation ability towards transition metal ions also enhances antioxidant properties of these compounds.

In [45], condensation of thiacalix[4]arene hydrazides in *1,3-alternate* **31a,b** and *cone* stereoisomeric forms **32a,b** with 3,5-di-*tert*-butyl-4-hydroxybenzaldehyde **33** was investigated. The result of these reactions depends on the spatial structure and conformational mobility of initial hydrazides. Condensation of tetrahydrazides **31a,b** with *1,3-alternate* stereoisomeric form of calix[4]arene platform give only tetrasubstituted hydrazones **34a,b**.

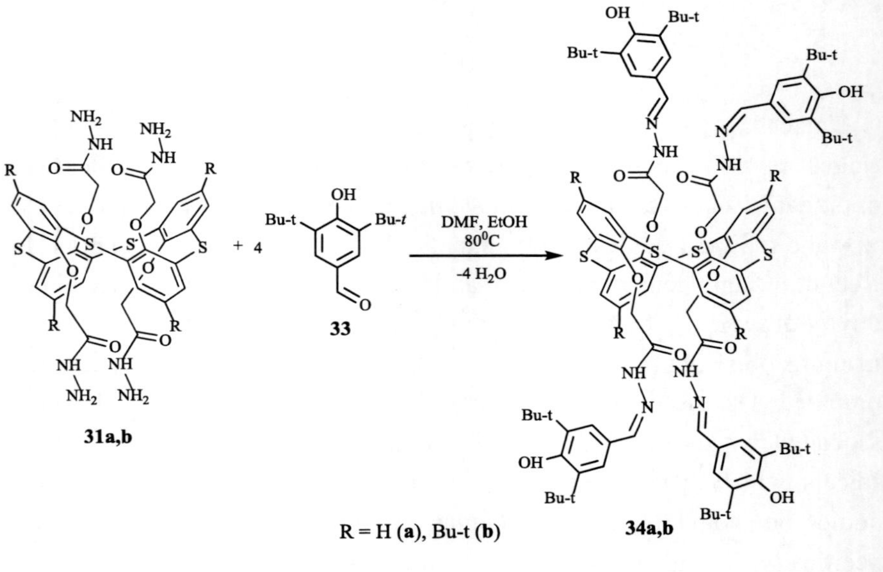

31a,b R = H (**a**), Bu-t (**b**) **34a,b**

The result of the reaction of cone-shaped tetrahydrazides **32a,b** is determined by the presence of *tert*-butyl groups on the upper rim of tetrathiacalix[4]arene. When there are no *tert*-butyl groups, only tetrahydrazone **35** is formed, while in their presence bishydrazone **36** with intramolecular N,N'-diacetylhydrazine bridge is formed. In the latter case, tetrahydrazone product was not detected in the reaction mixture.

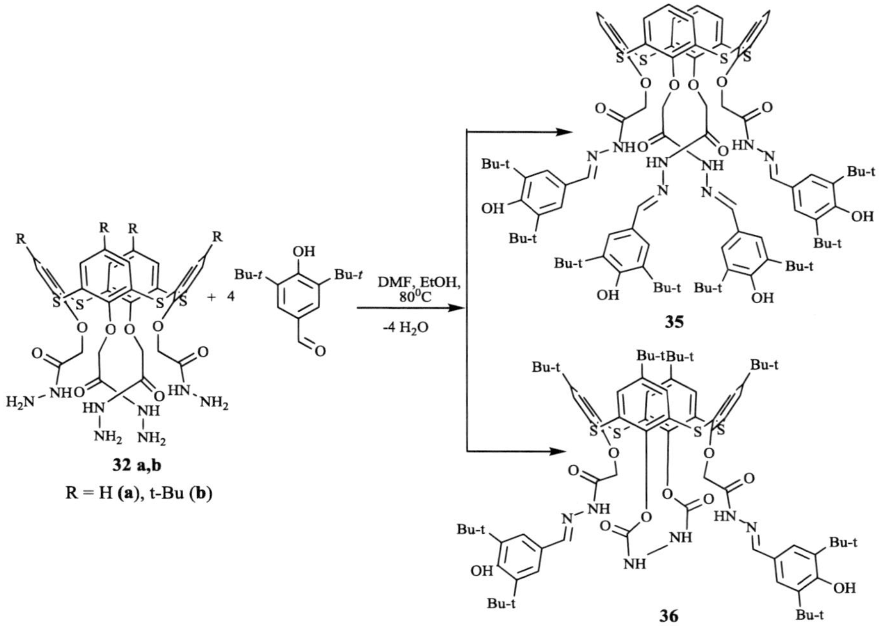

Analogous proceeding is observed during the reaction of hydrazide **32b** with benzaldehyde and 4-nitrobenzaldehyde [46]. The main reason of this unusual reaction for tetrahydrazide derivative of 4-*tert*-butyltetrathiacalix[4]arene **32b** in *cone* stereoisomeric form is presumably the mutual repulsion of bulky *tert*-butyl groups of the upper rim. For this reason, endo-oriented hydrazide groups of the lower rim approach each other, which facilitates their disproportionation and the formation of N,N'-diacetylhydrazine bridge.

Investigation of the complexation ability of compounds **35** and **34b** by liquid extraction showed that they are selective extractants of silver and mercury ions [47].

In [48], the synthesis of calix[n]arenes **37a,b** modified by unsymmetrical hindered phenolic fragments on the upper rim of macrocycle was described.

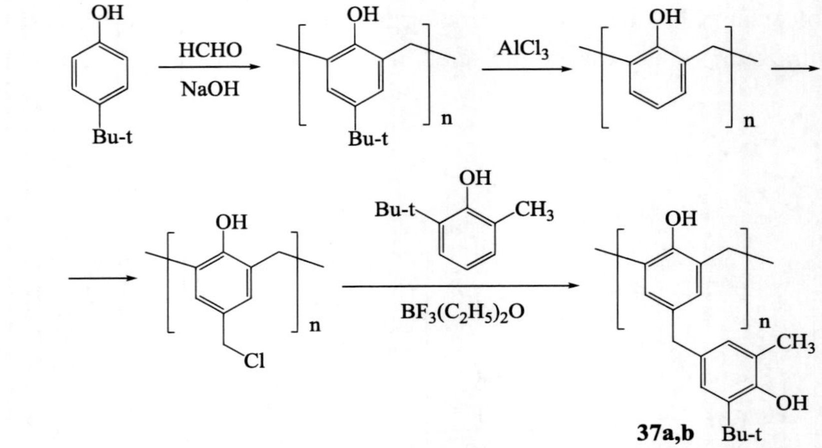

37a,b
n = 4 (**a**), 5 (**b**)

Compound **37a** was isolated in *cone* conformation. The authors from [49] investigated the mechanism of the formation of the triplet state and calix[4]aryloxy radical of compound **37a** by time-resolve laser flash photolysis.

4. Calix[n]resorcinols Modified by Hindered Phenolic Fragments

In [50-53], the reaction of *tetra*-alkylcalix[4]resorinols **21**, **25a,b** and **38a-e** with 3,5-di-*tert*-butyl-4-hydroxybenzylacetate **39** was studied.

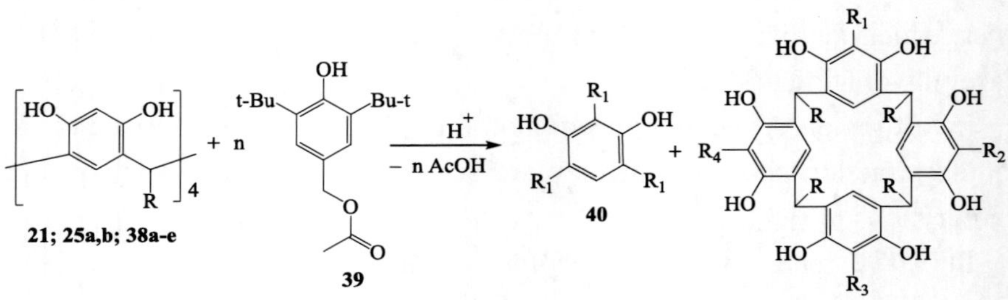

R=Me(**21,41h**), Et (**25a,41a**), Pr (**38a, 41c**), C_5H_{11} (**38b, 41d**), C_7H_{15} (**25b, 41b**), C_8H_{17} (**38c, 41e**), C_9H_{19} (**38d, 41f**), $C_{11}H_{23}$ (**38e,41g**);

$R^1=R^2=R^3=R^4=-CH_2-$ 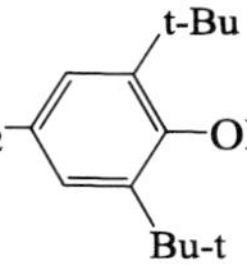

$-OH$ (**40, 41a,c,d,h**); $R^1=R^2$ ($R^1=R^3$) $=-CH_2-$ 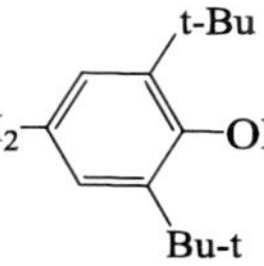

$-OH$

$R^3=R^4$ ($R^2=R^4$)=H (**41b,e,f**); $R^1=-CH_2-$ 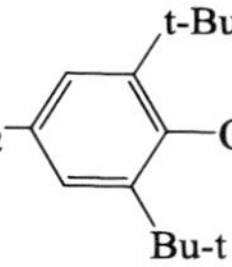

$-OH$; $R^2=R^3=R^4=H$ (**41g**)

Benzylation of calix[4]resorcinol **21** in formic acid occurs with the formation of the reaction mixture containing 80% of tetrabenzyl derivative **41h** and 20% of 2,4,6-tris(3,5-di-*tert*-butyl-4-hydroxybenzyl)resorcinol **40.** IR spectroscopy investigations of compound **41h** showed that hydroxylic groups of calix[4]resorcinol scaffold participate in "pseudocooperative" intramolecular hydrogen bonds (IHBs) of the upper rim and form additional IHBs of their protons with π-electrons near proximal benzene rings. The formation of the latter IHBs indicates the implementation of the predominant *cone* conformation [52]. The data of one- and two-dimensional NMR spectroscopy confirm this suggestion [51].

The formation of compound **40** in the reaction of calix[4]resorcinol **21** with benzyl acetate **39** is the first example of the decomposition of calix[4]resorcinol ring by electrophilic reagent under mild conditions. Because macrocycle does not destruct in the "calix[4]resorcinol **41h** – acid catalyst" binary system, the authors suggest that the formation of compound **40** is the result of decomposition of calix[4]resorcinols **21** and/or **41h** as the result of *ipso*-substitution by in situ generated 3,5-di-*tert*-butyl-4-hydroxybenzyl carbocation.

The reaction of tetraethyl, tetrapropyl, and tetrapentylcalix[4]-resorcinols **25a,c,d** and **38a,b** with benzyl acetate **39** in the presence of formic acid gives tetrabenzylated calix[4]resorcinols **41a,c,d** and low amounts of tris-benzylated resorcinol **40** [52]. Further increase in the size of alkyl substituents restricts benzylation reaction. The reaction of calix[4]resorcinols **25b** and **38c,d** containing heptyl, octyl, and nonyl fragments on the lower rim of molecule with benzyl acetate **39** affords compounds **41b**,**e**,**f** possessing only two 3,5-di-*tert*-butyl-4-hydroxybenzyl fragments [53]. Because there is a double increase in the number of signal of aromatic protons in ^{1}H NMR spectra of compounds **41b**,**e**,**f**, the authors consider that these compounds represent the mixture of two regioisomers.

With the transition to calix[4]resorcinol **38e** containing undecyl fragment on the lower rim of molecule monobenzylated calixresorcinol **41g** is formed as the only product in the reaction with benzyl acetate **39.** It should be noted that the products of macrocycle opening were not detected in the case of the formation of debenzylated and monobenzylated calix[4]resorcinols.

The synthetic results are presumably caused by the ability of calix[4]resorcinols to form organized structures [54]. With an increase in the length of alkyl radical, the ability of self-association increases and calix[4]resorcinols exist in denser packing in this case, which does not destruct even at high dilution. For this reason, the attack of bulky electrophilic particle represented by 3,5-di-*tert*-butyl-4-hydroxybenzyl carbocation is restricted, which influences the degree of benzylation of calixarene matrix.

The method of self-organization of calixarenes in solution depends on their structure, concentration, and nature of solvent. Among mentioned factors, the structure of calixarenes plays a decisive role. In the review from [56], the formation of polycapsules, nanotubes, and polycylinders with the participation of calixarene matrix was shown. The self-organization of amphiphilic calixarenes and calixresorcinols was studied in most detail [55-57]. Depending on the cavity size and conformation, these compounds may form aggregates with lamellar packing of molecules, conventional micelles, or monomolecular micelles.

In [53], self-organization of benzylated calix[4]resorcinols **41b**,**d**,**f**,**g**,**h** was studied by dielkometry and dynamic light scattering. Although, the mechanism of self-organization of the compounds under study is unclear, one can suggest with a sufficient reliability a significant role of geometrical factor during the formation of organized structures along with the energy component. So-called supramolecular polymers, that is, nanochains composed of several calixresorcinol fragments are presumably formed in the systems under study.

Tetrabenzylated calix[4]resorcinols **41a,c,d,h** were investigated in phosphorylation, silylation, and acetylation [52]. Phosphorylation of compounds **41a,c,d,h** with dichloromethylphosphonate gives rise to cavitands **42a-d**:

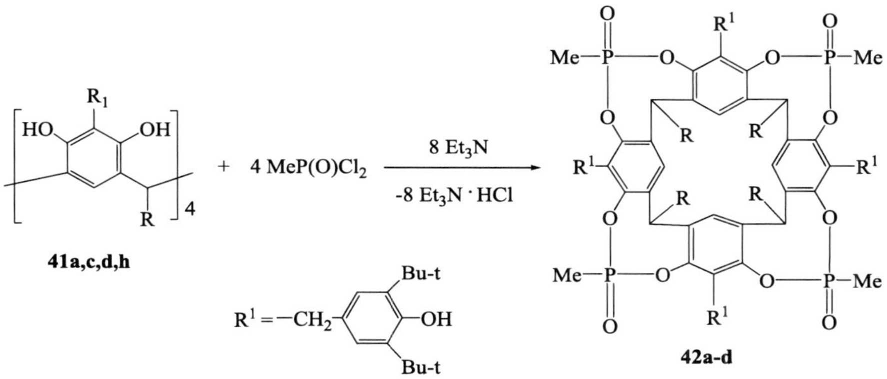

41a,c,d,h **42a-d**

R = Me (**41h, 42a**), Et (**41a, 42b**), Pr (**41c, 42c**), C_5H_{11} (**41d, 42d**).

Phosphorylated hindered phenolic derivatives of tetraalkylcalix[4]resorcinols **44a-d** were prepared in one step in the reaction of corresponding calix[4]resorcinols with phosphorylated methylene quinones **43a,b**. The reaction does not stop at the benzylation step of calix[4]resorcinol and is accompanied by dealkoxy(phenoxy)lation to give heterocycles [41]:

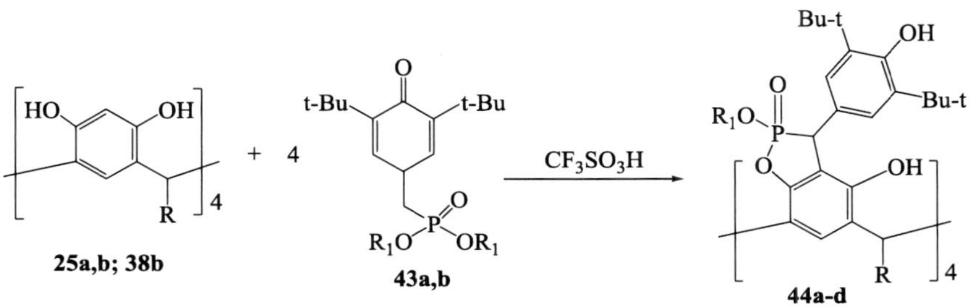

25a,b; 38b **43a,b** **44a-d**

R=Et (**25a, 44a, 44d**), C_5H_{11} (**38b, 44b**), C_7H_{15} (**25b, 44c**); R_1=Me (**43a, 44a-c**), Ph (**43b, 44d**)

During the reaction of calix[4]resorcinols **41a,c,d,h** with dimethylchlorosilane in toluene in the presence of triethylamine (24 h, 100°C) organosilicon cavitands **45a-d** are formed.

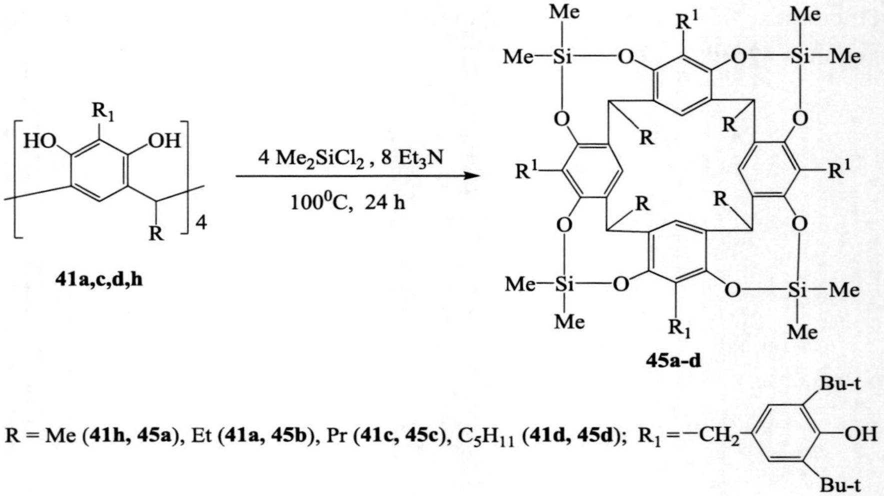

As a result of silylation of calix[4]resorcinols **41a,d,h** with hexamethyldisilazane, octasilyl derivatives **46a-c** were produced.

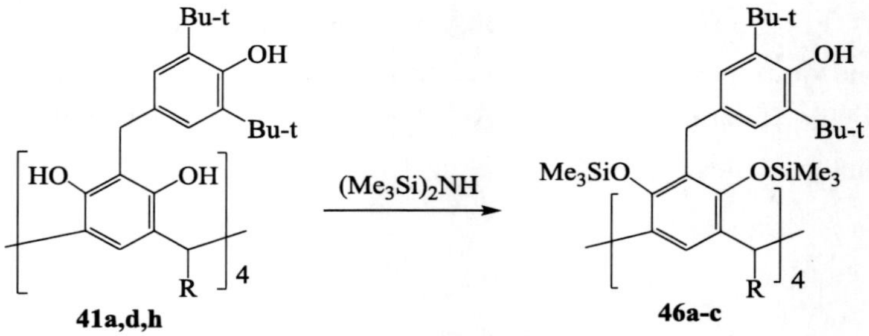

R = Me (**41h 46a**), Et (**41a, 46b**), C_5H_{11} (**41d, 46c**)

The reaction of calix[4]resorcinols **41a,c,d,h** with acetic anhydride in the presence of catalytic amounts of pyridine gives the products of partial or full acylation depending on reaction conditions:

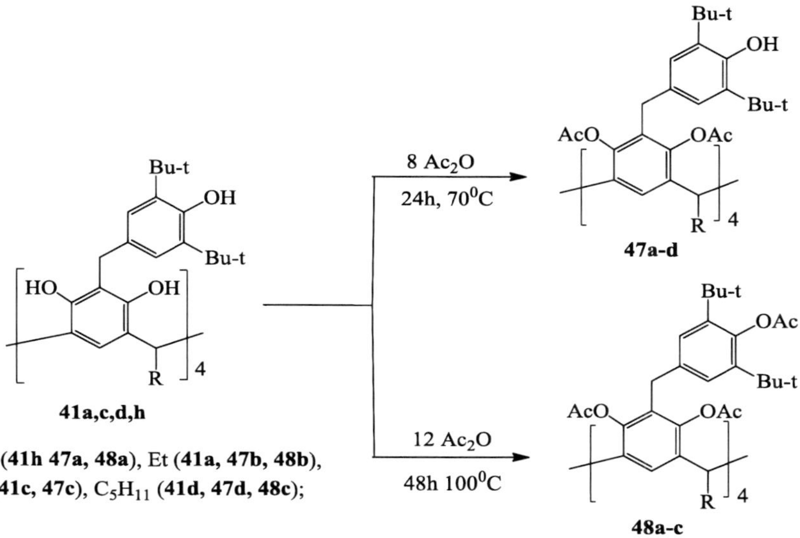

41a,c,d,h **47a-d** **48a-c**

R = Me (**41h 47a, 48a**), Et (**41a, 47b, 48b**), Pr (**41c, 47c**), C_5H_{11} (**41d, 47d, 48c**);

In [58], the effect of tetrabenzylated calix[4]resorcinol **41h** on thermal oxidation stability of low-density polyethylene (LDPE) and rubber compounds based on nitrile rubbers SNK-26 and SNK-18 was investigated. Under the conditions of processing and accelerated ageing of LDPE, calix[4]resorcinol **41h** is superior to benzylated resorcinol **40** and similar to high-performance industrial antioxidant Irganox 1010 according to antioxidant action.

In the rubber compound based on rubber SKN-18 with high filler content, benzylated calixarene **41h** is slightly superior to unmodified calixresorcinol **21** and amine stabilizer 4010NA by the retention of tensile strength and gives way to them according to the retention of elongation at break. In the rubber compound based on rubber SKN-26 with low filler content possessing higher physicomechanical parameters, calixresorcinol **41h** is more effective inhibitor of thermal oxidation destruction as compared to calixresorcinol **21** and comparable to amine stabilizer 4010NA. Thus, the modification of calixresorcinol **21** with 3,5-di-*tert*-butyl-4-hydroxybenzyl fragments is an approach to increase the antioxidant activity of this macrocyclic stabilizer.

In [59], the interaction of dialkylaminomethyltetraalkyl-calix[4]resorcinols **49a-f** with 3,5-di-*tert*-butyl-4-phenylmercaptan **50** was studied. Deamination of calix[4]resorcinols **49a-f** occurs under the action of mercaptan **50** with the formation of novel calix[4]resorcinols **51a,c,e,f** in *cone* stereoisomeric form modified by sulfur-containing hindered phenolic fragments in 63–72% yields.

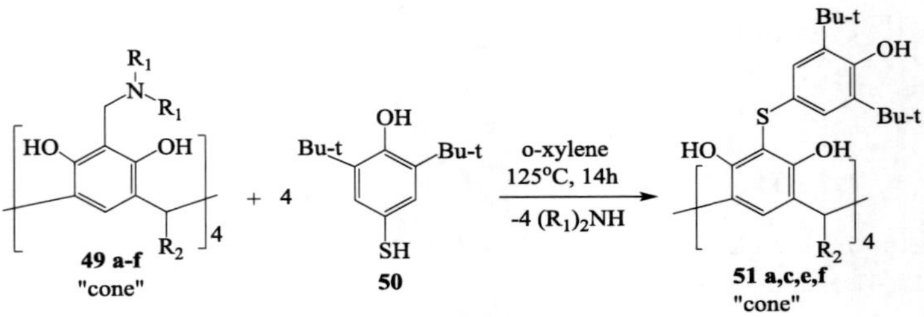

R_1=R_2=CH_3 (**a**); R_1=C_2H_5, R_2=CH_3 (**b**); R_1=CH_3, R_2=C_2H_5 (**c**); R_1=R_2=C_2H_5 (**d**); R_1=CH_3, R_2=C_7H_{15} (**e**); R_1=CH_3, R_2=C_9H_{19} (**f**)

It should be noted that, in contrast with the described reactions of tetraalkylcalix[4]resorcinols with benzyl acetate **39**, tetrasubstituted derivatives containing four hindered phenolic fragments in molecule were isolated in all cases.

An interesting feature of the synthesis of calix[4]resorcinol **51a** from the corresponding dimethylaminomethyl derivative **49a** involves that this compound is formed only when using freshly synthesized dialkylaminomethylated calix[4]resorcinol **49a**. It is known [60] that calix[4]resorcinol molecules **49a** form dimers in the crystal after removal of solvent, where dimethylaminomethyl groups of neighboring molecules are included in macrocycle cavities of each other and, consequently, are sterically inaccessible for nucleophilic attack.

Therefore, calix[4]resorcinol **49a** loses its reactivity in the reaction with 3,5-di-*tert*-butyl-4-phenylmercaptan **50** upon storage (that is, with the evaporation of solvent molecules scavenged by the calixarene matrix). When using calixarene **49a** stored for 3 months, the mixture of the products of incomplete substitution of dimethylamino groups is formed in the reaction mixture according to ^{1}H NMR spectroscopy data even after more than 60-h reflux. Washing of calix[4]resorcinol **49a**, which was stored for a long time, with dimethylsulfoxide and, then, water restore its activity. That is, the introduction of solvent molecules destructs the dimeric structure of compound **49a**.

The use of diethylaminomethylcalix[4]resorcinol **49b** in the reaction with 3,5-di-*tert*-butyl-4-phenylmercaptan **50** is not related to the mentioned effect, because the molecules of calixresorcinol **49b** do not form mentioned dimers in the crystal; rather they possess another supramolecular organization [60], where dimethylaminomethyl groups are available for nucleophilic attack.

Antioxidant activity of calix[4]resorcinols **51a,e,f** was studied in the model reaction of initiated oxidation of styrene. It was determined that compound **51a** inhibits styrene oxidation to the higher extent than compounds **51e,f** and industrial sulfur-containing stabilizer represented by bis-(3,5-di-*tert*-butyl-4-hydroxybenzyl)sulfide.

Antioxidant activity of calix[4]resorcinols **51a,c,e,f** decreases with an increase in the length of hydrocarbon radical on the lower rim. This may presumably be caused by self-association of molecules **51a,c,e,f**, which are similar to those described above in a series of 3,5-di-*tert*-butyl-4-hydroxybenzylated calix[4]resorcinols. These processes may lead to the spatial shielding of antioxidant phenolic fragments. It should be noted that even in the presence of such unfavorable effect, calix[4]resorcinols **51e,f** are superior to industrial sulfur-containing hindered phenolic antioxidant bis-(3,5-di-*tert*-butyl-4-hydroxybenzyl)sulfide according to antioxidant activity, which proves the perspective of the employment of calix[4]resorcinol platform for the synthesis of effective polyfunctional antioxidants.

Another approach to the synthesis of hindered phenolic derivatives of calix[4]resorcinols, which is alternative to the modification of the calixresorcinol matrix, is the introduction of hindered phenolic fragments during the synthesis of macrocycle. Condensation of resorcinol and its derivatives **52a-d** with hydroxybenzaldehydes **53a,b** catalyzed by concentrated hydrochloric acid in ethanol affords calix[4]resorcinols **54a-h** bearing four hindered phenolic fragments on the lower rim of macrocycle in 36–50% yields [61].

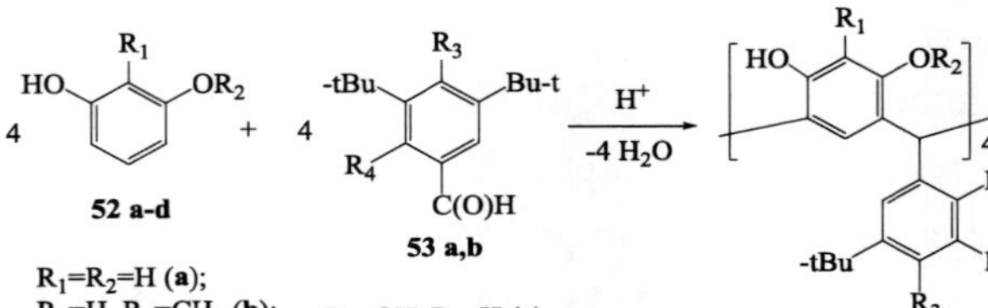

52 a-d

R_1=R_2=H (**a**);
R_1=H, R_2=CH_3 (**b**);
R_1=CH_3, R_2=H (**c**);
R_1=OH, R_2=H (**d**)

53 a,b

R_3=OH, R_4=H (**a**);
R_3=H, R_4=OH (**b**)

54 a-h

R_1=R_4=H, R_2=R_3=OH (**a**);
R_1=R_3=H, R_2=R_4=OH (**b**);
R_1=R_4=H, R_2=OCH_3, R_3=OH (**c**);
R_1=R_3=H, R_2=OCH_3, R_4=OH (**d**);
R_1=CH_3, R_2=R_3=OH, R_4=H (**e**);
R_1=CH_3, R_2=R_4=OH, R_3=H (**f**);
R_1=R_3=OH, R_2=R_4=H (**g**);
R_1=R_4=OH, R_2=R_3=H (**h**)

Catalysis of this reaction by trifluoromethanesulfonic acid in nonpolar 1,4-dioxane solvent provides an increase in the yield of compound **56a** up to 80%. According to X-ray analysis, compounds **56a,b,c** exist in *chair* stereoisomeric form in the crystal (*rctt*, C_{2h}-symmetry). Doubling of signals of the protons of resorcinol rings in ^{1}H NMR spectra proves the retention of this space form in solution.

CONCLUSION

In review, the results of synthesis and properties of antioxidants based on calixarenes and calixresorcinols have been summarized.

The authors have synthesized for the first time tetraalkylcalix[4]-resorcinols containing 3,5-di-*tert*-butyl-4-hydroxybenzyl and 3,5-di-*tert*-butyl-4-hydroxythiobenzyl fragments on the upper rim of macrocycle. The effect of supramolecular organization of calix[4]-resorcinols on their reactivity has been shown. Using condensation of resorcinol and its derivatives with hydroxybenzaldehydes, vinyl phosphonates, aminoacetals, the synthesis of calix[4]resorcinols with hindered phenolic fragments on the lower rim of macrocycle has been carried out for the first time.

First representative of thiacalix[4]arenes in *cone* and *1,3-alternate* stereoisomeric forms containing acylhydrazone fragments of hindered hydroxybenzaldehydes have been synthesized. It has been determined that these calixarenes are selective extractants of silver and mercury ions. High antioxidizing activity of the synthesized compounds has been demonstrated by the example of thermal oxidation of polyethylene and nujol oil, model reaction of initiated oxidation of styrene, and thermal ageing of butadiene-nitrile rubbers.

REFERENCES

[1] Podyachev, S.N.; Sudakova, S.N.; Syakaev, V.V.; Galiev, A.K.; Shagidullin, R.R.; Konovalov A.I. The preorganization effect of the calix[4]arene platform on the extraction properties of acetylhydrazide groups with transition metal ions. *Supramol. Chem.*, 2008, *20* (5), 479-486.

[2] Podyachev, S.N.; Kashapova, N.E.; Syakaev, V.V.; Sudakova, S.N.; Zainullina, R.R.; Gruner, M.; Habicher, W.D.; Barsukova, T.A.; Yang, F.; Konovalov, A.I. Mercury(II) and silver(I) receptors based on tatrathiacalix[4]arene hydrazones. *J. Incl. Phenom. Macrocycl. Chem.*, 2014, *78* (1-4), 371-380.

[3] de Fatima, A.; Fernandes, S.A.; Sabino, A.A. Calixarenes as New Platforms for Drug Design. *Curr. Drug Discov. Technol.*, 2009, *6* (2), 151-170.

[4] Burton, L.P.J. Phenolic Phosphite Antioxidant for Polymers. Pat. US 4474917, October 2, 1984.

[5] Ehrhardt, D.; Hauptman, S.; Mann, G.; Mertens, W.; Noll, B.; Weinelt, F.; Weinelt, H. Cyclische mehrkernphenole. Pat. DE 4022920 A1, June 20, 1991.

[6] Pastor, S.D.; Odorisio, P. Acylated Calixarene Stabilizers. Pat. US 4617336, October 14, 1986.

[7] Seiffarth, K.; Schulz, M.; Goermar, G.; Bachmann, J. Calix[n]arene - new light stabilizers for polyolefins. *Polym. Degrad. and Stab.*, 1989, *24*, 73-80.

[8] Seiffarth, K.; Goermar, G.; Haubold, W.; Roehr, Ch.; Schulz, M.; Knopel, R. Verfahren zur Stabilisierung von Hochpolymeren, niedermolekularen organischen Verbindungen und Olen. Pat. DD 264446, February 1, 1989.

[9] Feng, W.; Yuan, L.H.; Zheng, S.Y.; Huang, G.L.; Qiao, J.L.; Zhou Y. The Effect of p-*tert*-Butylcalix[n]arene on γ-Radiation Degradation of Polypropylene. *Rad. Phys. and Chem.,* 2000, *57* (3-6), 425-429.

[10] Jipa, S.; Zaharescu, T.; Setnescu, R.; Dumitru, M.; Gorghiu, L.M.; Mihalcea, I.; Bumbac M. Effect of calixarenes on thermal stability of polyethylenes. *Polym. Degrad. and Stab.*, 2003, *80* (2), 203-208.

[11] Gorghiu, L.M.; Dumitrescu, C.; Olteanu, R.L.; Bumbac, M.; Jipa, S. Comparative study of some calix[n]arenes and other phenolic compounds influence on the thermal stability of LDPE. *Romanian Reports in Physics*, 2004, *56* (3), 466-472.

[12] Ohkatsu, Y.; Nishiyama, T. Phenolic antioxidants-effect of ortho-substituents. *Polym. Degrad. and Stab*., 2000, *67* (2), 313-318.

[13] Ohkatsu, Y.; Ishikawa, S.; Tobita E. Consideration on the effect of ortho-substituents of phenols by semiempirical molecular orbital method MOPAC. *Polym. Degrad. and Stab.*, 2000, *67* (3), 541-545.

[14] Matsuura, T.; Ohkatsu, Y. Phenolic antioxidants: effect of o-benzyl substituents. *Polym. Degrad. and Stab.*, 2000, *70* (2), 59-63.

[15] Kajiyama, T.; Ohkatsu, Y. Effect of para- substituents of phenolic antioxidants. *Polym. Degrad. and Stab*., 2001, *71* (3), 445-452.

[16] Asfari, Z.; Bohmer, V.; Harrowfield, J.; Vicens, J. (Eds.). *Calixarenes 2001,* Dordrecht, Kluwer Academic Press., 2001.

[17] Perrin, R.; Lamartine, R.; Perrin, M. The potential industrial applications of calixarenes. *Pure and Appl. Chem.*, 1993, *65* (7), 1549-1559.

[18] Chennakesavulu, K. Raviathul, B.M.; Bhaskar, R.G.; Prabhakar, S. Study on thermal decomposition of calyx[6]arene and calyx[8]arene. *J. Therm. Anal. Calorim*., 2011, *103* (3), 853-862.

[19] Seiffarth, K.; Schulz, M.; Goermar, G.; Bachmann, J. Stabilisatoren für niedermolekulare und hochpolymere organische Materialien. Pat. DD 273844 B5, November 29, 1989.

[20] Sase S., Yoshimura Y., Suzuki H. Pat. 62227988. Japan (1986). *C.A.*, 1988, *108*, 113563.

[21] Kadir, Z.A.; Yoshii, F.; Makuuchi, K.; Ishigaki I. Durability of Radiation-Sterilized Polymers. XII. The Effects of Nucleating Agent on the Oxidative Degradation of Polypropylene. *Polymer*, 1989, *30* (8), P. 1425-1432.

[22] Yuan, L.H.; Feng, W.; Zheng, S.Y.; Huang, G.L. The Effect of p-tert-Butylcalixarene on Radiation Degradation of Polypropylene. *Journal of Radiation Research and Radiation Processing*, 1997, *4*, 25-31.

[23] Feng, W.; Yuan, L.H.; Zheng, S.Y.; Huang, G.L.; Qiao, J.L. The Study of p-tert-Butylcalix[n]arene on y-Radiation Stabilization of Polypropylene. *Proceedings of the 8th Japan-China Bilateral Symposium on Radiation Chemistry*, Kyodai Kaikan, Kyoto, Japan, October 25 – 29, 1999.

[24] Zaharescu, T.; Jipa, S.; Setnescu, R.; Santos, C.; Gigante, B.; Mihalcea, I.; Podină, C.; Gorghiu, L. Thermal stability of additivated isotactic polypropylene. *Polym. Bull.*, 2002, *49* (4), 289-296.

[25] Jipa, S.; Zaharescu, T.; Setnescu, R.; Setnescu, T.; Dumitru, M.; Gorghiu, L.; Mihalcea I. Efectele de stabilizare induse în polipropilenă de antioxidanţi cu funcţiuni fenolice. *Materiale Plastice*, 2002, *39*, 160-164.

[26] Goermar, G.; Haubold, W.; Rohr, Ch.; Taplick, T.; Seiffarth, K.; Kocke, A. Verfahren zur Stabilisierung von Hochpolymeren niedermolekularen organischen Verbindungen und Olen. Pat. DD 278140, April 25, 1990.

[27] Yang, W.; Kambara, H.; Osawa, M.; Ueno, Y.; Okoshi, M. Resin Additive Composition and Process for Producing the Same. Pat. US 2003/0057406, May 27, 2003.

[28] Yordanova, S.; Miloshev, S. Thermal investigation of calixarene-containing PVC composites. *Journal of University of Chemical Technology and Metallurgy*, 2006, *41* (4), 397-402.

[29] Consoli, G.M.L.; Galante, E.; Daquino, C.; Granata, G.; Cunsolo, F.; Geraci C. Hydroxycinnamic acid clustered by a calixarene platform: radical scavenging and antioxidant activity. *Tetrahedron Lett.*, 2006, *47* (37), 6611-6614.

[30] Erdem, E.; Öztürk, R.; Aydin, M.; Mamadov, R.; Söyleyici S. Antioxidant and DPPH (1,1-diphenyl-2-picrylhydrazyl) Free Radical Scavenging Activities of Boniger Acid and Calix[4]arene Derivative. *International Journal of Secondary Metabolite*, 2014, *1* (1), 21-22.

[31] Seiffarth, K.; Schulz, M.; Goermar, G.; Bachmann, J. Stabilisatoren fur niedermolekulare und hochpolymere organische materialien. Pat. DD 254739 A1, March 09, 1988.

[32] Böhmer, V.; Dörrenbächer, R; Frings, M.; Heydenreich, M.; de Paoli, D.; Vogt, W.; Ferguson, G.; Thondorf, I. Annelated Calixarenes Composed of Calix[4]arenes with Hydroxy Groups in the Endo- and Exo Position. *J. Org. Chem.*, 1996, *61*, 549-559.

[33] Ehrhardt, D. Calix[4]arenpolyole – Eine neue Gruppe von Alterungsschutzmitteln fur Elastomere. Part I. *Gummi, Fasern Kunststoffe*, 1992, *45* (5), 231-239.

[34] Ehrhardt D. Calix[4]arenpolyole – Eine neue Gruppe von Alterungsschutzmitteln fur Elastomere. Part II. *Gummi, Fasern Kunststoffe*, 1992, *45* (7), 358-361.

[35] Ehrhardt, D.; Hauptmann, S.; Mann, G.; Mertens, W.; Weinelt, F.; Noll, B.; Weinelt, H. Antioxidants for natural and synthetic rubbers Pat. DD 290427, May 29, 1991.

[36] Ehrhardt, D.; Hauptmann, S.; Mann, G.; Mertens, W.; Weinelt, F.; Noll, B.; Weinelt, H. Phenolic calixarenes as stabilizers for natural and synthetic rabbers. Pat. DD 291902, July 18, 1991.

[37] Hasbullah, A.; Abosadiya, H.A.; Jumina, J.; Tahir, M.; Yamin, B.M. Synthesis, Structural and Antioxidant Properties of C-p-methoxyphenylcalix[4]resorcinarene. *International Journal on Advanced Science, Engineering and Information Technology*, 2013, *3* (2), 36–39.

[38] Mehrer, M. Neue Lichtschutzmittel auf Basis von Calix[n]arenen und Resorcin[n]arenen. Pat. EP 0896980 A1, February 17, 1999.

[39] Vovk, A.I.; Shivanyuk, A.M.; Bugas, R.V.; Muzychka, O.V.; Melnyk, A.K. Antioxidant and antiradical activities of resorcinarene tetranitroxides. *Bioorg. Med. Chem. Lett.*, 2009, *19* (5), 1314-1317.

[40] Görmar, G.; Seiffarth, K.; Wendt, D.; Burtzlaff, C. Calix[n]arenderivate mit antioxidativen Wirkungen. *J. Prakt. Chem.*, 1990, *332* (1), 37-42.

[41] Kasymova, E.M.; Shaekhov, T.R.; Burilov, A.R.; Pudovik, M.A.; Konovalov, A.I. New phosphorylated calix[4]resorcinoles containing the sterically hindered phenol. *Books of Abstracts,* V-th International Symposium "Supramolecular Systems in Chemistry and biology," Kyiv, Ukraine, May 12-16, 2009.

[42] Storozhok, N.M.; Burilov, V.A.; Daryukhina, N.Yu.; Borisov, A.I.; Veselov, E.N.; Daryukhina, E.N.; Gureeva, N.V. The antioxidant properties of calixarenes (in Russian). *Butlerov Communications*, 2016, *46* (4), 15-22.

[43] Voigt, J. *Die Stabilisierung der Kunststoffe gegen Licht und Wärme*. Berlin: Springer, 1966.

[44] Khizhan, E.I.; Vinogradov, V.V.; Morenko, V.V.; Nikolaevskii, A.N.; Khizhan, A.I.; Zarechnaya, O.M.; Dmitruk, A.F. Antiradical activity of aryl- and hetarylhydrazones in ethylbenzene oxidation. *Russ. J. Gen. Chem.*, 2013, *83* (8), 1529–1536.

[45] Podyachev, S.N.; Barsukova, T.A.; Bukharov, S.V.; Syakaev, V.V.; Sudakova, S.N.; Nugumanova, G.N.; Konovalov, A.I. Synthesis of hydrazones containing sterically hindered phenol groups and tetrathiacalix[4]arene fragment. *Russ. J. Org. Chem.*, 2012, *48* (10), 1360-1365.

[46] Podyachev, S.N.; Sudakova, S.N.; Syakaev, V.V.; Shagidullin, R.R.; Galiev, A.K.; Konovalov, A.I. Synthesis of tetrathiacalix4.arene functionalised by acetylhydrazone fragments. *Mendeleev Commun.*, 2006, *6*, 297-299.

[47] Podyachev, S.N.; Burmakina, N.E.; Sudakova, S.N.; Syakaev, V.V.; Konovalov, A.I. Improvement of selective d-cation binding by tetrathiacalix[4]arene hydrazides: synthesis and extraction properties. *Supramol. Chem.*, 2010, *22* (6), 339-346.

[48] Tu, H.Y.; Yuan, L.H.; Jiang, J.K.; Huang, Sh.; Tian, B.Z. Synthesis of Unsymmetrically Hindered Phenol-Substituted Calixarenes. *Chin. J. Org. Chem.*, 2000, *20* (5), 836-839.

[49] Feng, W.; Yuan, L.; Yao, S.; Wang W. Nanosecond Flash Photolysis of Unsymmetrical Phenol-Substituted Calix[4]Arene in Cyclohexane. Proceedings of the 8th Japan-China bilateral symposium on radiation chemistry, Kyodai Kaikan, Kyoto, Japan, October 25-29, 1999.

[50] Bukharov, S.V.; Nugumanova, G.N.; Mukmeneva, N.A.; Syakaev, V.V.; Burilov, A.R.; Pudovik, M.A.; Konovalov, A.I. Unusual reaction of tetramethylcalix[4]resorcinolarene with 3,5-di-*tert*-butyl-4-hydroxybenzyl acetate. *Russ. J. Gen. Chem.*, 2003, *72* (8), 1320-1321.

[51] Bukharov, S.V.; Nugumanova, G.N.; Mukmeneva, N.A.; Teregulova, E.A.; Burilov, A.R.; Pudovik, M.A.; Nikolaeva, I.L.; Kasymova, E.M.; Konovalov, A.I. Reaction of Tetramethylcalix[4] resorcinolarene with 3,5-Di-*tert*-butyl-4-hydroxybenzyl Acetate. *Russ. J. Org. Chem.*, 2003, *39* (5), 689-691.

[52] Kasymova, E.M.; Burilov, A.R.; Mukmeneva, N.A.; Bukharov, S.V.; Nugumanova, G.N.; Pudovik, M.A.; Chernova, A.V.; Shagidullin, R.R.; Konovalov, A.I. Synthesis and Some Properties of Tetrakis-3,5-di-*tert*-butyl-4-hydroxybenzylated Calix[4]resorcinols. *Russ. J. Gen. Chem.*, 2007, *77* (3), 458-468.

[53] Kasymova, E.M.; Burilov, A.R.; Mukmeneva, N.A.; Bukharov, S.V.; Nugumanova, G.N.; Ibragimova, A.R.; Timosheva, A.P.; Zakharova, L.Ya.; Kudryavtseva, L.A.; Pudovik, M.A.; Konovalov, A.I. Benzylation of tetraalkylcalix[4]resorcinols. *Russ. Chem. Bull.*, 2009, *58* (1), 138-144.

[54] Brunsveld, L.; Folmer, B.J.B.; Meijer, E.W.; Sijbesma, R.P. Supramolecular Polymers. *Chem. Rev.*, 2001, *101*, 4071-4097.

[55] Shinkai, S.; Arimura, T.; Araki, K. Kawabata H. Syntheses and aggregation properties of new water-soluble calixarenes. *J. Chem. Soc. Perkin Trans. 1*, 1989, *11*, 2039-2045.

[56] Arimori, S.; Nagasaki, T.; Shinkai, S. Self-assembly of tetracationic amphiphiles bearing a calix[4]arene core. Correlation between the

core structure and the aggregation properties. *J. Chem. Soc., Perkin Trans. 2*, 1995, *4*, 679-683.
[57] Rebek, J. Host–guest chemistry of calixarene capsules. *J. Chem. Com.*, 2000, *8*, 637-643.
[58] Bukharov, S.V.; Teregulova, E.A.; Nugumanova, G.N.; Mukmeneva, N.A.; Miryasova, F.K.; Burilov, A.R.; Pudovik, M.A.; Nikolaeva, I.L.; Kasymova, E.M.; Konovalov, A.I. Calix[4]resorcinols as stabilizers for rubber stocks based on butadiene-acrylonitrile rubbers. *Russ. J. Appl. Chem.*, 2003, *76* (11), 1867-1869.
[59] Nugumanova, G.N.; Barsukova, T.A.; Bukharov, S.V.; Mukmeneva, N.A.; Burilov, A.R.; Syakaev, V.V. Synthesis of 3,5-di-*tert*-butyl-4-hydroxyphenylsulfanylmethyl-substituted Tetramethylcalix[4]re-sorcinarenes. *Russ. J. Org. Chem.*, 2010, *46* (7), 1097-1098.
[60] Gubaidullin, A.T.; Nikolaeva, I.L.; Kharitonov, D.I.; Litvinov, I.A.; Bashmakova, N.I.; Burilov, A.R.; Pudovik, M.A.; Konovalov, A.I. Effect of substituents on the supramolecular structure and stability of crystals of dialkylaminomethylated calix[4]resorcinolarenes. *Russ. J. Gen. Chem.*, 2002, *72* (2), 259-267.
[61] Barsukova, T.A.; Bukharov, S.V.; Burilov, A.R.; Nugumanova, G.N.; Syakaev, V.V.; Mukmeneva, N.A. Synthesis of new calyx[4]resorcines with sterically hindered phenol fragments (in Russian). *Butlerov Communications*, 2011, *28* (17), 48-51.

BIOGRAPHICAL SKETCHES

Sergey Bukharov

Affiliation: Kazan National Research Technological University of the Ministry of Education and Science of Russia

Education: Full Professor

Research and Professional Experience: Synthesis, chemistry and technology of antioxidants of polymers, lubricating oils, fuels and bioantioxidants. Synthesis and properties of polyfunctional compounds based on hindered phenols and hetero- and macrocycles.

Professional Appointments: Head of the Department of General Organic and Petrochemical Synthesis Technology of Kazan National Research Technological University

Publications from the Last 3 Years:

1) Chugunova, E.; Akylbekov, N.; Shakirova, L.; Dobrynin, A.; Syakaev, V.; Latypov, Sh.; Bukharov, S.; Burilov, A. Synthesis of hybrids of benzofuroxan and N-, S-containing sterically hindered phenols derivatives. Tautomerism. *Tetrahedron*, 2016, *72*, 6415-6420.
2) Tagasheva, R.G.; Ibatullina, E.D.; Bukharov, S.V.; Burilov, A.R.; Krivolapov, D.B.; Nugumanova, G.N. Azo coupling of 2-hydroxymethyl-4-*tert*-butylphenol with sulfamoylbenzenediazonium chloride. *Russ. J. Gen. Chem.*, 2016, *86* (3), 764-766.
3) Tagasheva, R.G.; Bukharov, S.V.; Nugumanova, G.N.; Oludina, Y.N.; Sazykina, M.A.; Sazykin, I.S.; Seliverstov, E.Y.; Khammami, M.I. Synthesis and antioxidant activity of modified sterically hindered phenols. *Russ. J. Gen. Chem.*, 2016, *86* (3), 602-606.
4) Bukharov, S.V.; Burilov, A.R.; Oludina, Y.N.; Tagasheva, R.G.; Syakaev, V.V.; Musin, R.Z.; Nugumanova, G.N. Synthesis of phosphorylated derivatives of isatin with sterically hindered phenol fragments. *Phosphorus, Sulfur Silicon Relat. Elem.*, 2016, *191* (7), 1069-1074.
5) Nugumanova, G.N.; Barsukov, T.A.; Bukharov, S.V.; Syakaev, V.V.; Deberdeev, R.Y. Synthesis and antioxidant activity of 3,5-di-*tert*-butyl-4-hydroxyphenylthiomethyltetraalkylcalix[4]resorcinarenes. *Russ. J. Gen. Chem.*, 2015, *85* (9), 2087-2090.

6) Oludina, Y.N.; Sazykina, M.A.; Prazdnova, E.V.; Sazykin, I.S.; Khmelevtsova, L.E.; Bukharov, S.V.; Barsukova, T.A.; Tagasheva, R.G.; Nugumanova, G.N. Synthesis of Modified Sterically Hindered Phenols and Investigation of Their Ability to Protect Bacterial DNA Against UVB Damage. *Pharm. Chem. J.*, 2015, *48* (12), 787-790.
7) Oludina, Yu.N.; Ibatullina, E.D.; Bukharov, S.V.; Nugumanova, G.N.; Tagasheva, R.G.; Deberdeev, R.Ya. Reaction of alkyl splitting of 4-acetylamino-N-(3,5-di-*tert*-butyl-4-hydroxybenzyl)-benzene-sulfonamide. *Russ. J. Gen. Chem.*, 2015, *85* (2), 383-386.
8) Nugumanova, G.N.; Bukharov, S.V.; Tagasheva, R.G.; Mukmeneva, N.A.; Deberdeev, R.Y. Antioxidant activity of isatin acylhydrazones with sterically hindered phenol fragments. *Russ. J. Gen. Chem.*, 2015, *85* (1), 53-56.
9) Barsukova, T.A.; Nugumanova, G.N.; Bukharov, S.V.; Podyachev, S.N.; Syakaev, V.V.; Mukmeneva, N.A.; Deberdeev, R.Ya. Synthesis of acylhydrazones of sterically hindered hydroxybenzaldehydes based on phloroglucinol. *Russ. J. Org. Chem.*, 2014, *50* (6), 805-809.
10) Oludina, Yu.N.; Bukharov, S.V.; Burilov, A.R.; Tagasheva, R.G.; Syakaev, V.V.; Musin, R.Z.; Akhmetova, E.F.; Nugumanova, G.N. Synthesis of sterically hindered phenol-containing phosphorylated isatin derivatives. *Russ. Chem. Bull.*, 2014, *63* (1), 115-117.
11) Oludina, Yu.N.; Voloshina, A.D.; Kulik, M.V.; Zobov, V.V.; Bukharov, S.V.; Tagasheva, R.G.; Nugumanova, G.N.; Burilov, A.R.; Kravchenko, M.A.; Skornyakov, S.N.; Rusinov, G.L. Synthesis, toxicity, and antituberculosis activity of isoniazid derivatives containing sterically hindered phenols. *Pharm. Chem. J.*, 2014, *48* (1), 5-7.

Gulnara Nugumanova

Affiliation: Kazan National Research Technological University of the Ministry of Education and Science of Russia

Education: Associate Professor

Research and Professional Experience: Research in the field of degradation and stabilization of polymers, fuels and oils. Synthesis, properties and technology of N, P, S-containing antioxidants with hindered phenolic fragments.

Professional Appointments: Assistant Professor of the Department of Synthetic Rubber Technology of Kazan National Research Technological University

Publications from the Last 3 Years:

1) Tagasheva, R.G.; Ibatullina, E.D.; Bukharov, S.V.; Burilov, A.R.; Krivolapov, D.B.; Nugumanova, G.N. Azo coupling of 2-hydroxymethyl-4-*tert*-butylphenol with sulfamoylbenzenediazonium chloride. *Russ. J. Gen. Chem.*, 2016, *86* (3), 764-766.
2) Tagasheva, R.G.; Bukharov, S.V.; Nugumanova, G.N.; Oludina, Y.N.; Sazykina, M.A.; Sazykin, I.S.; Seliverstov, E.Y.; Khammami, M.I. Synthesis and antioxidant activity of modified sterically hindered phenols. *Russ. J. Gen. Chem.*, 2016, *86* (3), 602-606.
3) Bukharov, S.V.; Burilov, A.R.; Oludina, Y.N.; Tagasheva, R.G.; Syakaev, V.V.; Musin, R.Z.; Nugumanova, G.N. Synthesis of phosphorylated derivatives of isatin with sterically hindered phenol fragments. *Phosphorus, Sulfur Silicon Relat. Elem.*, 2016, *191* (7), 1069-1074.
4) Nugumanova, G.N.; Barsukov, T.A.; Bukharov, S.V.; Syakaev, V.V.; Deberdeev, R.Y. Synthesis and antioxidant activity of 3,5-di-*tert*-butyl-4-hydroxyphenylthiomethyltetraalkylcalix[4]resorcinarenes. *Russ. J. Gen. Chem.*, 2015, *85* (9), 2087-2090.
5) Oludina, Y.N.; Sazykina, M.A.; Prazdnova, E.V.; Sazykin, I.S.; Khmelevtsova, L.E.; Bukharov, S.V.; Barsukova, T.A.; Tagasheva, R.G.; Nugumanova, G.N. Synthesis of Modified Sterically Hindered

Phenols and Investigation of Their Ability to Protect Bacterial DNA Against UVB Damage. *Pharm. Chem. J.*, 2015, *48* (12), 787-790.
6) Oludina, Yu.N.; Ibatullina, E.D.; Bukharov, S.V.; Nugumanova, G.N.; Tagasheva, R.G.; Deberdeev, R.Ya. Reaction of alkyl splitting of 4-acetylamino-N-(3,5-di-*tert*-butyl-4-hydroxybenzyl)-benzene-sulfonamide. *Russ. J. Gen. Chem.*, 2015, *85* (2), 383-386.
7) Nugumanova, G.N.; Bukharov, S.V.; Tagasheva, R.G.; Mukmeneva, N.A.; Deberdeev, R.Y. Antioxidant activity of isatin acylhydrazones with sterically hindered phenol fragments. *Russ. J. Gen. Chem.*, 2015, *85* (1), 53-56.
8) Barsukova, T.A.; Nugumanova, G.N.; Bukharov, S.V.; Podyachev, S.N.; Syakaev, V.V.; Mukmeneva, N.A.; Deberdeev, R.Ya. Synthesis of acylhydrazones of sterically hindered hydroxybenzaldehydes based on phloroglucinol. *Russ. J. Org. Chem.*, 2014, *50* (6), 805-809.
9) Oludina, Yu.N.; Bukharov, S.V.; Burilov, A.R.; Tagasheva, R.G.; Syakaev, V.V.; Musin, R.Z.; Akhmetova, E.F.; Nugumanova, G.N. Synthesis of sterically hindered phenol-containing phosphorylated isatin derivatives. *Russ. Chem. Bull.*, 2014, *63* (1), 115-117.
10) Oludina, Yu.N.; Voloshina, A.D.; Kulik, M.V.; Zobov, V.V.; Bukharov, S.V.; Tagasheva, R.G.; Nugumanova, G.N.; Burilov, A.R.; Kravchenko, M.A.; Skornyakov, S.N.; Rusinov, G.L. Synthesis, toxicity, and antituberculosis activity of isoniazid derivatives containing sterically hindered phenols. *Pharm. Chem. J.*, 2014, *48* (1), 5-7.

Alexander Burilov

Affiliation: A. E. Arbuzov Institute of Organic and Physical Chemistry of Kazan Scientific Center of Russian Academy of Sciences

Education: Full Professor

Research and Professional Experience: Investigation of the reactions of various P, N-functionally substituted α, β, γ-aminoacetals with phenols, polyphenols, calixarenes, allowing the synthesis of various linear, heterocyclic, macrocyclic compounds (including functionalized calixarenes) containing aryl and organophosphorus fragments in the structure.

The development of new, generic approaches to the synthesis of the original multivalent phosphorus, nitrogen-containing ligands on the calix [4] resorcinol platform and linear polyphenols – intermediates for the design of nanocomposites. The development of a «one-pot» synthesis method for previously unknown framework compounds containing a nodal phosphonate group, consisting in the condensation of 2-ethoxyvinyl dichlorophosphonate with a variety of phenols (phenol, hydroquinone, α-, β-naphthols, resorcinol, 2-methylresorcinol, pyrogallol).

Synthesis of new FOS-bioantioxidants, as well as broad-spectrum biologically active substances containing in the composition hindered phenolic fragments, furoxane groups.

Professional Appointments: Head of the laboratory of Organoelemental Synthesis named after A.N. Pudovik A. E. Arbuzov Institute of Organic and Physical Chemistry of Kazan Scientific Center of Russian Academy of Sciences

Honors: The title of "Honored Worker of Science of the Republic of Tatarstan"

Publications from the Last 3 Years:

1) Pashirova, T.N.; Gibadullina, E.M.; Burilov, A.R.; Kashapov, R.R.; Zhiltsova, E.P.; Syakaev, V.V.; Habicher, W.D.; Rummeli, M.H.; Latypov, Sh.K.; Konovalov, A.I.; Zakharova, L.Ya. Amphiphilic O-functionalized calix[4]resocinarenes with tunable structural behavior. *RSC Adv*., 2014, *4*, 9912–9919.

2) Chugunova, E.A.; Voloshina, A.D.; Mukhamatdinova, R.E.; Serkov, I.V.; Proshin, A.N.; Gibadullina, E.M.; Burilov, A.R.; Kulik, N.V.; Zobov, V.V.; Krivolapov, D.B.; Dobrynin, A.B.; Goumont, R. The Study of the Biological Activity of Amino-Substituted Benzofuroxans. *Letters in Drug Design & Discovery*, 2014, *11* (4), 502-512.
3) Sadykova, Yu.M.; Dalmatova, N.V.; Nagimova, M.F; Burilov, A.R.; Pudovik, M.A.; Sinyashin, O.G. Novel naphtol based bicyclic phosphonates. *Heteroatom. Chem.*, 2014, *25* (1), 55-59.
4) Knyazeva, I.R.; Matveeva, V.I.; Syakaev, V.V.; Gabidullin, B.M.; Gubaidullin, A.T.; Gruner, M.; Habicher, W.D.; Burilov, A.R.; Pudovik, M.A. Thiophosphorylated derivatives of metha- and ortho-hydroxybenzaldehydes in one step synthesis of novelcalix[4]resorcinols. *Tetrahedron Lett.* 2014, *55* (52), 7209-7214.
5) Gazizov, A.S.; Smolobochkin, A.V.; Voronina, Yu. K.; Burilov, A.R.; Pudovik, M.A. New methods of synthesis of 2-arylpirrolidines: reaction of resorcinol and its derivatives with γ-ureidoacetals. *ARKIVOC*, 2014, *IV*, 319-327.
6) Vagapova, L.I.; Voronina, Yu. K.; Syakaev, V.V.; Sharafutdinova, D.R.; Amirova, L.R.; Burilov, A.R.; Pudovik, M.A.; Garyfzyanov, A.R.; Sinyashin, O.G. Phosphorylated Aminoacetals in Synthesis of New Acyclic, Cyclic, and Heterocyclic Polyphenol Structures. *Heteroatom. Chem.*, 2014, *25* (3), 178-185.
7) Chugunova, E.; Boga, C.; Sazykin, I.; Cino, S.; Micheletti, G.; Mazzanti, A.; Sazykina, M.; Burilov, A.; Khmelevtsova, L.; Kostina, N. Synthesis and antimicrobial activity of novel structural hybrids of benzofuroxan and benzothiazole derivatives. *Eur. J. Med. Chem.*, 2015, *93*, 349-359.
8) Sadykova, Y.M.; Dalmatova, N.V.; Voronina, Y.K.; Burilov, A.R.; Pudovik, M.A.; Sinyashin, O.G.; Mitrasov, Y.N. Modification of bicyclic phocphonates on resorcinol and its derivatives via the Mannich reaction. *Heteroatom Chem.*, 2015, *26* (3), 224-230.

9) Kibardina, L.K.; Gazizov, A.S.; Burilov, A.R.; Pudovik, M.A.; Chumakova, L.V. Reaction of pyridoxal with phenols: synthesis of novel 1-aryl-substituted furopyridines. *Synthesis*, 2015, *47* (5), 721-725.
10) Gazizov, A.S.; Smolobochkin, A.V.; Voronina, J.K.; Burilov, A.R.; Pudovik, M.A. Acid-Catalyzed Reaction of (4,4-Diethoxybutyl)ureas with Phenols as a Novel Approach to the Synthesis of α-Arylpyrrolidines. *Synth. Comm.*, 2015, *45* (1), 1215-1221.
11) Gazizov, A.S.; Kharitonova, N.I.; Smolobochkin, A.V.; Syakaev, V.V.; Burilov, A.R.; Pudovik, M.A. Facile synthesis of 2-(2-arylpyrrolidin-1-yl)pyrimidines via acid-catalyzed reaction of N-(4,4-diethoxybutyl)pyrimidin-2-amine with phenols. *Monatshefte fur Chemie*, 2015, *146* (11), 1845-1849.
12) Gazizov, A.S.; Smolobochkin, A.V.; Voronina, J.K.; Burilov, A.R.; Pudovik, M.A. Acid-catalyzed ring opening in 2-(2-hydroxynaphthalene-1-yl)-pyrrolidine-1-carboxamides: a pathway to the synthesis of dibenzoxanthenes, diarylmethanes and calixarenes. *Tetrahedron*, 2015, *71* (3), 445-450.
13) Khrizanforova, V.V.; Knyazeva, I.R.; Sokolova Matveeva, V.I.; Nizameev, I.R.; Gryaznova, T.V.; Kadirov, M.K.; Burilov, A.R.; Sinyashin, O.G.; Budnikova, Yu.H. Nickel complexes of thiophosphorylated calix[4]resorcinols as effective catalysts for hydrogen evolution. *Electrocatalysis*, 2015, *6* (4), 357-364.
14) Knyazeva, I.R.; Sokolova, V.I.; Sharafutdinova, D.R.; Burilov, A.R.; Pudovik, M.A.; Gazizov, M.B. Synthesis of novel phosphorus-containing calix[4]resorcinols based on polyphenols with phosphonium fragment. *Heteroatom. Chem.*, 2015, *26* (1), 1-4.
15) Smolobochkin, A.V.; Gazizov, A.S.; Burilov, A.R.; Pudovik, M.A.; Khuzyasheva, D.G.; Amirov, R.R. Synthesis of 2-arylpyrrolidines by reaction of γ- ureidoacetals with benzene-1,3,5-triol. *Russ. J. Org. Chem.*, 2016, *52*, 538–540.

16) Kadirov, M.K.; Knyazeva, I.R.; Nizameev, I.R.; Safiullin, R.A.; Matveeva, V.I.; Kholin, K.V.; Sinyashin, O.G. Oxygen reduction reaction catalyzed by nickel complexes based on thiophosphorylated calix[4]resorcinols and immobilized in the membrane electrode assembly of fuel cells. *Dalton Trans.*, 2016, *45*, 16157–16161
17) Gazizov, A.S.; Kharitonova, N.I.; Syakaev, V.V.; Dobrynin, A.B.; Burilov, A.R.; Pudovik, M.A. Synthesis of benzooxadiazocines via the acid-catalyzed reaction of pyrimidine-containing acetals with resorcinol derivatives. *Monatshefte fur Chemie*, 2016, *147*, 2113-2117.
18) Chugunova, E.A.; Akylbekov, N.I.; Voloshina, A.D.; Kulik, N.V.; Zobov, V.V.; Babaev, V.M.; Gavrilov, N.V.; Burilov, A.R. Synthesis and investigation of antimicrobial activity of compounds derived from benzo[C][1,2,5]oxadiazole-1-oxides and phenolates. *Synth. Comm.*, 2016, *46* (18), 1560-1565
19) Chugunova, E.; Mukhamatdinova, R.; Sazykina, M.; Dobrynin, A.; Sazykin, I.; Karpenko, A.; Mirina, E.; Zhuravleva, M.; Gavrilov, N.; Karchava, Sh.; Burilov, A. Synthesis of new "Hybrid" Compounds based on Benzofuroxans and Aminoalkylnaphthalimides. *Chemical Biology and Drug Design*, 2016, *87* (4), 626-634.
20) Chugunova, E.; Akylbekov, N.; Bulatova, A.; Gavrilov, N.; Voloshina, A.; Kulik, N.; Zobov, V.; Dobrynin, A.; Syakaev, V.; Burilov, A. Synthesis and biological evaluation of novel structural hybrids of benzofuroxan derivatives and fluoroquinolones. *Europ. J. Med. Chem.*, 2016, *116*, 165-172.
21) Chugunova, E.; Akylbekov, N.; Bulatova, A.; Gavrilov, N.; Voloshina, A.; Kulik, N.; Zobov, V.; Dobrynin, A.; Syakaev, V.; Burilov, A. Synthesis and biological evaluation of novel structural hybrids of benzofuroxan derivatives and fluoroquinolones. *Europ. J. Med. Chem.*, 2016, *116*, 165-172.

22) Azmukhanova, R.R.; Gibadullina, E.M.; Nguyen, T.T.; Burilov, A.R.; Pudovick, M.A. Novel nitrogen-, phosphorus-containing sterically hindered phenols. *Phosphorus, Sulfur Silicon Relat. Elem.*, 2016, *191* (11-12), 1457-1458.

23) Amirova, L.R.; Burilov, A.R.; Amirova, L.M.; Bauer, I.; Habicher, W.D. Kinetics and mechanistic investigation of epoxy-anhydride compositions cured with quaternary phosphonium salts as accelerators. *J. Polym. Sci. Pol. Chem.*, 2016, *54* (3), 1088-1097.

INDEX

B

C

D

E

F

G

H

I

K

L

M

Q

R

S

T

U

V

W

X

Z